**Applied Engineering
Mechanics**

MECHANICAL ENGINEERING

A Series of Textbooks and Reference Books

EDITORS

L. L. FAULKNER
Department of Mechanical Engineering
The Ohio State University
Columbus, Ohio

S. B. MENKES
Department of Mechanical Engineering
The City College of the
City University of New York
New York, New York

OTHER VOLUMES IN PREPARATION

Applied Engineering Mechanics

STATICS AND DYNAMICS

G. Boothroyd and C. Poli
University of Massachusetts
Amherst, Massachusetts

MARCEL DEKKER, INC.   New York and Basel

Library of Congress Cataloging in Publication Data

Boothroyd, Geoffrey, [Date]
    Applied engineering mechanics.

    (Mechanical engineering ; v. 5)
    Includes index.
    1. Mechanics, Applied. I. Poli, C., [Date]
joint author. II. Title. III. Series.
TA350.B5398   620.1   80-19044
ISBN 0-8247-6945-7

MARCEL DEKKER, INC.
270 Madison Avenue, New York, New York 10016

Current printing (last digit):
10 9 8 7 6 5 4 3 2 1

PRINTED IN THE UNITED STATES OF AMERICA

To Shirley and Ann

Preface

This book is intended primarily for those students following engineering or technology programs in which a knowledge of the principles of engineering mechanics (statics and dynamics) is required.

Most present texts dealing with engineering mechanics emphasize the use of vector notation. This is mathematically convenient and sound when solving three-dimensional problems. However, it is felt by the present authors that an introduction to mechanics should emphasize two-dimensional problems for which vector notation is of little advantage. In fact, the use of vector notation can prevent the student from obtaining the desirable physical "feel" for engineering situations. In addition, the great majority of practical engineering mechanics problems are two-dimensional in nature; for example, three-dimensional mechanisms are most rare, and three-dimensional aerospace problems form only a very small percentage of all engineering situations. In presently available texts, artificial three-dimensional problems are introduced simply to serve as a vehicle for the vector mathematics. Indeed, these problems are often examples of bad design practice and will prevent the student from obtaining an appreciation of good design.

The authors also abhor the current attitude in education that a text on a given subject should deal only with that subject and not encroach on other subjects. This attitude is typified by the teacher ignoring spelling mistakes and poor grammar in the report of an engineering experiment. The students will argue that separate courses exist for the teaching of English composition and that all that matters in an engineering experiment is that the student has successfully completed the experiment. This attitude seems to have found its way into texts on engineering mechanics. Pictorial representations of engineering situations are now the norm rather than engineering drawings. It appears that, through the lack of proper college and high school education, engineering students cannot be expected to understand engineering drawings and hence must be shown pictures—preferably three-dimensional and multicolored. It is the authors' view that students must be able to understand engineering-type drawings, the principles of projection, etc., and that the more practice they get the better. For this reason, most of the artwork in the present text is in the form of simple line drawings. Further, the students should learn to translate a word description of an engineering situation without the aid of a picture. For this reason, many problems in the later chapters of the present text are word problems.

This text is divided into two major areas: statics and dynamics. In addition, particularly in that portion of the book dealing with statics, the chapters are divided so that the first few deal primarily with the theory (Chapters 1–5 for statics and Chapters 12–19 for dynamics), while the remaining chapters deal primarily with the application of the theory to problems of engineering significance (Chapters 6–11 for statics and Chapters 20 and 21 for dynamics). This division of the material differs from that typically found in most of today's texts dealing with mechanics. With the approach presented here, the student first concentrates on learning the basic underlying principles and applying them to rather simplified problems; the student then moves on to the solution of some practical engineering problems which make use of various combinations of the basic principles. This approach is used because it has been the authors' experience that while students seem able to solve certain types of problems when they appear in a chapter which emphasizes the particular principle or principles encountered, they are often unable to cope with these same types of problems when they are divorced from a particular chapter in the text. Thus, the present approach helps the student develop the confidence necessary to tackle problems that are not tied to a particular chapter in a textbook.

Every attempt is made in this book to follow the latest International Standards Organization (ISO) recommendations for units and definitions. The most important of these is, of course, the International (SI) System of basic and derived units. The basic system is described in ISO Standard 1000, and recommendations for its application are specified in ISO Recommendation R31. The latter document includes standard symbols for the various derived units. This book is written entirely in SI units because the United States is already in the process of conversion to the SI metric system.

Both the approach used here and the class notes from which this book has been developed have been classroom-tested several times at the University of Massachusetts at Amherst and have met with the enthusiastic approval of the students. It is the authors' feeling that in using the present text students will learn about a variety of aspects of engineering practice in addition to the theory of engineering mechanics.

Finally, we would like to thank Ms. Donna Mougin and Ms. Jan Peene for typing the manuscript and Ms. Lyla Wilson for preparing the illustrations.

G. Boothroyd
C. Poli

Contents

**Applied Engineering
Mechanics**

1

Some Basic Concepts of Mechanics

1.1 Introduction

Mechanics is a physical science that can be subdivided into *statics*, which deals with the forces acting on bodies that are at rest (in static equilibrium), and *dynamics*, which deals with bodies in motion. Dynamics is further subdivided into *kinematics*, which deals with the motion of bodies without regard to the forces causing the motion, and *kinetics*, which deals with the relationships between the forces acting on a body and its resulting motion.

In this text no consideration is given to the deformation of a body or the tendency of a body to deform under the action of the forces applied. Such considerations would fall within the subjects of *dynamics of fluids* and *mechanics of deformable bodies* or *mechanics of materials*.

The fundamental quantities used in mechanics are length, time, and mass. All other variables used in analysis can be derived from these. Thus, it is necessary first to define these quantities, then to obtain the derived quantities, and finally to express, in mathematical terms, the basic relationships among the derived quantities. These basic relationships are obtained from fundamental laws commonly known as Newton's laws of motion. These laws are really deductions based on experiment and physical observations.

1.2 Fundamental Quantities of Length, Time, and Mass

Length is a measure of displacement or relative position. In ancient times, the *forearm* (cubit) was used as the standard unit of length. However, since forearms differ in size from person to person, obvious difficulties arise in

1

using such a definition. Later, in 1793, the French used the length of a straight line scratched on a bar kept under closely monitored conditions in Paris as the standard unit of length. The length of the line was called a metre and was one ten-millionth of the distance from the equator to the north pole on a line running through Paris. In 1889 the definition of the metre was standardized as the distance (at 0°C) between two fine lines on a platinum-iridium bar preserved at the International Bureau of Weights and Measures in Sevres, France. Today, the metre is defined* as follows: "The metre is the length equal to 1,650,763.73 wavelengths in vacuo of the radiation corresponding to the transition between the levels $2p_{10}$ and $5d_5$ of the krypton-86 atom." This standard is believed to be reproducible to about 2 parts in 10^8. This standard is not used for everyday work. National Standards laboratories will calibrate reference standards up to 1 metre by direct interferometric methods and to an accuracy of about 1 part in 10^7. These reference standards are used to calibrate the various working standards used in industry.

Mass is a measure of the amount of material in a body. The standard reference mass (and not of weight or of force) of 1 kilogram is held in Paris in the form of a solid platinum-iridium cylindrical block. The reference mass was legalized at the 1st CGPM held in 1889. To compare masses, a balance scale must be used.

Time is a measure of the succession of events. Originally the unit of time, the second, was defined as the fraction 1/86,400 of the mean solar day. The exact definition of "mean solar day" was left to astronomers, but their measurements have shown that, because of irregularities in the rotation of the earth, the use of the mean solar day does not provide the desired accuracy. The difficulty with this early definition was that one could not measure a second by direct comparison with the interval of time defining the second; instead, lengthy astronomical observations were needed. To define the unit of time more precisely, the 11th CGPM (1960) adopted a definition, given by the International Astronomical Union, which was based on the tropical year. Experimental work, however, had already shown that an atomic standard of time interval, based on a transition between two energy levels of an atom or a molecule, could be realized and reproduced much more accurately. Considering that a very precise definition of the unit of time of the International System, the second, is indispensable for the needs of advanced metrology, the 13th CGPM (1967) decided to replace the definition of the second by the

*The definition of length was defined at the 11th General Conference of Weights and Measures (CGPM) in 1960. For a more detailed discussion of this definition, one may consult "The International System of Units (SI)," *National Bureau of Standards Special Publication* **330**, for sale by the Superintendent of Documents, U.S. Government Printing Office, Washington, D.C. 20402.

following: "The second is the duration of 9,192,631,770 periods of the radiation corresponding to the transition between the two hyperfine levels of the ground state of the cesium-133 atom."

1.3 Derived Quantities: Velocity and Acceleration

The general study of the relationships between length and time is called *kinematics*. Figure 1.1a shows an object represented by the point P which is moving along a straight path denoted by the line Os. At a certain time t_1 the object is located a distance s_1 from the stationary reference point O, and at a later time t_2 the object is located a distance s_2 from point O. The time interval under consideration is, therefore, $t_2 - t_1$ and can be represented by the symbol Δt. Thus, Δt means "the time interval." Similarly, the displacement of the object is $s_2 - s_1$, or Δs.

The *velocity* of the object is defined as the rate of change of its position; its average velocity v_{av} during the time interval under consideration can be obtained by dividing the displacement by the time interval. Thus,

$$v_{av} = \frac{\Delta s}{\Delta t} \tag{1.1}$$

If the average velocity is measured over an infinitesimally short time interval

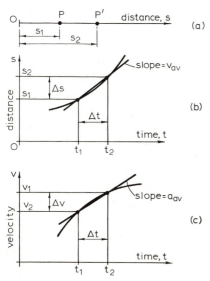

Fig. 1.1 Relationship among distance, velocity, and time in rectilinear motion.

(i.e., at a given instant) so that Δt approaches zero ($\Delta t \to 0$), then it is known as the instantaneous velocity v and at time t is given by $\Delta s/\Delta t$ in the limit as Δt approaches zero and is designated ds/dt.

In mathematical shorthand,

$$v = \lim_{\Delta t \to 0} \frac{\Delta s}{\Delta t} = \frac{d(s)}{dt} = \frac{ds}{dt} = \dot{s} \qquad (1.2)$$

Figure 1.1b illustrates this graphically, and it can be seen that, when Δt approaches zero, the instantaneous velocity v is the slope of the graph of distance plotted on a base of time t at the instant under consideration.

The instantaneous *acceleration* a of the object is defined as the rate of change of its velocity. Thus,

$$a = \lim_{\Delta t \to 0} \frac{\Delta v}{\Delta t} = \frac{d(v)}{dt} = \frac{dv}{dt} = \dot{v} \qquad (1.3)$$

Now, since $v = ds/dt$,

$$a = \frac{d^2 s}{dt^2} = \ddot{s} \qquad (1.4)$$

Figure 1.1c illustrates that, when Δt approaches zero, the instantaneous acceleration is equal to the slope of the graph of velocity v plotted on a base of time t.

1.4 Newton's Laws of Motion and the Derived Quantity of Force

The study of the motion of material objects having finite mass is known as dynamics, and a convenient concept in this work is the *particle*. A particle is a material body of finite mass but having infinitesimally small dimensions with respect to the distances or lengths involved in defining its position or motion. Thus, for example, when studying the motion of the earth about the sun, the earth can be treated as a particle since the diameter of the earth is about 13×10^6 metres (8000 miles), while the distance from the earth to the sun is about 150×10^9 metres (93,000,000 miles).

If a particle is stationary or moving with constant velocity, it will continue in that state unless acted upon by an external force. This is one way of expressing *Newton's first law of motion* and introduces the concept of force. Intuitively, it is known that an object must be pushed or pulled if it is to be moved from rest. This push or pull is the externally applied force.

Force is defined in terms of mass in *Newton's second law of motion*, which can be stated as follows: "A particle acted upon by a force will move with an acceleration proportional to and in the direction of the force." The ratio of the force F and the acceleration a is constant for a given particle. This constant ratio is the mass m of the particle.

Thus,

$$\frac{F}{a} = m \tag{1.5}$$

or
$$F = ma \tag{1.6}$$

which is an expression of Newton's second law of motion in mathematical terms. It can now be seen that the first law is a special case of the second because if no force acts on the particle, it will remain at rest or continue to move with a constant velocity. The *unit of force* is the *newton* and is defined as *the force required to give a 1 kilogram mass an acceleration of 1 metre per second squared.*

Newton's third law of motion states that for every action or force there is an equal and opposite reaction. This means that if an object is pushed with a force, then the object pushes back with an equal and opposite force.

As a result of his interest in astronomy, Newton also formulated the univeral law of gravitation.* This law allows the mutual force of attraction between two particles (Fig. 1.2) to be calculated and can be expressed mathematically as follows:

$$F = \frac{Gm_1m_2}{r^2} \tag{1.7}$$

where F is the mutual force of attraction, m_1 and m_2 are the masses of the two particles, r is the distance between the particles, and G is the universal gravitational constant. The value of G was first determined experimentally by Henry Cavendish in 1797 by measuring the force of attraction between two lead balls; its value is now established as 6.67×10^{-11} m^3/kg·s^2.

Using this equation, it is possible to calculate the magnitude of the gravitational force of attraction exerted by the earth on an object resting on its surface. This force of attraction is defined as the object's *weight* F_w. In this calculation, the earth can be regarded as a particle having the same mass of 5.976×20^{24} kg and located at the earth's center. Thus, if the mass m_e of the earth is substituted for m_1, the mass m of the object is substituted for m_2, and

*This law was formulated by Newton based on the three laws of planetary motion enunciated by Kepler, who, in turn, derived the laws from the observation of the motions of planets made by Tycho Brahe.

Fig. 1.2   Force of attraction between two particles.

the radius of the earth r_e (6.371 × 10^6 m) is substituted for r, we get

$$F_w = \left(\frac{Gm_e}{r_e^2}\right)m$$

$$= 9.824m \tag{1.8}$$

This means that an object placed on the earth's surface at a radius of 6.371 × 10^6 m from the center will be attracted to the center of the earth with a force given by Eq. (1.8) and is the weight of the object. The number 9.824 has the units of acceleration and is the acceleration with which the body would fall toward the earth's center if it were placed in a vacuum (to eliminate air resistance) and not supported.

The magnitude of this acceleration (which is given the symbol g) varies because the earth is not a perfect sphere. However, for an object on the earth's surface at a latitude of 45° the measured value is 9.807 m/s^2, and this value is the international standard. Thus,

$$F_w = mg \tag{1.9}$$

where $g = 9.807$ m/s^2.

In 1866, by an act of Congress, a pound mass (1 lbm) was defined as 0.4536 kg. Since a one pound force (1 lbf) is defined as the weight of a one pound mass at a point where the acceleration due to gravity is 9.81 m/s^2, from Eq. (1.9),

$$1 \text{ lbf} = 0.4536 \text{ (kg)} \times 9.81 \text{ (m/s}^2) = 4.45 \text{ kg·m/s}^2$$

or $1 \text{ lbf} = 4.45 \text{ N}$

1.5 The International (SI) System of Units

In 1960, the General Conference of Weights and Measures formally approved the system of units known as the International System of Units (SI, Systeme Internationale). This system has now been adopted throughout most of the world and is currently being adopted by the United States. In practice, the system is convenient because it obviates the need for the insertion of conversion factors into equations and eliminates many of the ambiguities present in other systems. The basic SI quantities of length, mass, and time have already been introduced, together with some of the derived quantities.

Appendix I is a summary of the quantities used in mechanics which will be introduced and used throughout this book. The ISO (International Standards Organization) recommendations for the symbols used for these quantities are also presented. In the SI system, prefixes are used to describe multiples and submultiples of units. For example, the *kilo* in kilogram means 1000 or 10^3.

Table 1.1 Prefixes Used with SI Units.

Amount	Multiple	Prefix	Symbol
1 000 000 000	10^9	giga	G
1 000 000	10^6	mega	M
1 000	10^3	kilo	k
0.001	10^{-3}	milli	m
0.000 001	10^{-6}	micro	μ
0.000 000 001	10^{-9}	nano	n

This prefix applied to the newton, i.e., kilonewton, would mean 1000 N. The prefixes presented in Table 1.1 are those most commonly used. It should be noted that

1. The prefix refers to the unit symbol; for example, the m (milli) in mm^3 means $(mm)^3$, not $m(m^3)$. A dot should be used to indicate multiplication when confusion could arise, i.e., $m \cdot m^3$.
2. Prefixes in denominators should be avoided, except for k in kg.

Example 1.1 Convert the units of the following measurements by applying the appropriate prefixes: (a) 5678 m, (b) 0.005 m, (c) 5×10^7 N.

Solution. (a) 5678 m = 5.678×10^3 m = 5.678 km, (b) 0.005 m = 5×10^{-3} m = 5 mm, (c) 5×10^7 N = 50×10^6 N = 50 MN.

Example 1.2 If an object travels 10 km in 5 Ms, what is its average velocity?

Solution. Average velocity = $\dfrac{\text{distance}}{\text{time}}$

$$= \frac{10 \times 10^3}{5 \times 10^6} = 2 \times 10^{-3} \text{ m/s}$$

$$= 2 \text{ mm/s}$$

1.6 Some Additional Basic Concepts and Definitions

It was explained earlier that many problems in mechanics can be solved by assuming that the object under consideration behaves as a particle of finite mass but of negligible dimensions. For other problems the concept of a rigid body is employed. A *rigid body* is one that does not deform under the action of forces. Although all bodies deform to some extent when forces are applied, the body can be assumed rigid when the difference between the initial and deformed configurations is negligible. Thus, during the launch phase of a rocket the deformation of the rocket may be neglected in order to determine the trajectory of the rocket and its orientation relative to the surface of the earth. On the other hand, if the problem is one of determining the internal deformations in the rocket structure due to the forces involved in launching, then the rocket cannot be assumed rigid.

In all problems in mechanics it is necessary to have a frame of reference from which measurements such as distances and angles can be measured. The basic frame of reference within which Newton's laws of motion are valid is known as an *inertial reference frame*. This reference frame is assumed to be "fixed" in space. While a reference frame which is stationary relative to the surface of the earth is not truly an inertial reference frame, for most engineering problems dealing with machines and structures on the earth's surface, the motion of such a reference frame can be neglected. For calculations of the motion of an artificial earth satellite, the motion of a reference frame attached to the surface of the earth becomes important. In this situation, a reference frame located at the center of the earth but not rotating with it can generally be treated as an inertial reference frame. For a study of the motion of an interplanetary probe, a reference frame located at the center of the sun will usually suffice as an inertial reference system.

A quantity which has magnitude only is known as a *scalar* quantity. Examples of scalar quantities are time, speed, volume, density, mass, and temperature.

A quantity which has direction as well as magnitude, which obeys the parallelogram law of addition, and which is commutative is a *vector*. Examples of vector quantities are displacement, velocity, acceleration, force, and moment. The time rate of change of vector quantities is affected by a change in either the direction or the magnitude of the vector quantity.

A vector quantity may be represented by a line having a length proportional to the magnitude of the quantity and having the same direction. Figure 1.3 shows an example where it will be noted that the direction of a vector is represented by an arrow. The arrow used to represent a vector quantity is called a vector. Mathematically, this vector can be represented by one of the symbols $\overline{oa}$, $\overline{oa}$, or **oa**. The order of the letters represents the direction of the arrow, and thus **oa** = −**ao**. Generally, it is more convenient to represent a vector quantity, such as force, by a single letter such as $\overline{F}$, or **F**.

Figure 1.4a shows two vectors **F**$_1$ and **F**$_2$. If these two vectors are added, the result is the vector **F**$_r$. That is,

$$\mathbf{F}_1 + \mathbf{F}_2 = \mathbf{F}_r \tag{1.10}$$

This can be illustrated by assuming that a person, starting at *o*, walks to *a* and

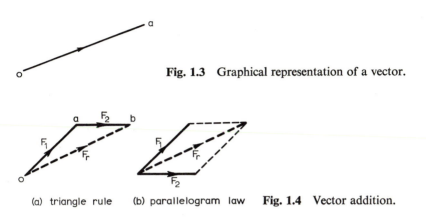

Fig. 1.3   Graphical representation of a vector.

(a) triangle rule   (b) parallelogram law   **Fig. 1.4**   Vector addition.

then to *b*. The result is the same as if the person had walked directly from *o* to *b*. Figure 1.4a illustrates the triangle rule for the addition of two vectors.

Figure 1.4b shows that the vectors can be added in any order. Thus, we can also write

$$F_2 + F_1 = F_r \qquad (1.11)$$

This is the commutative property of vectors. It can be seen that Fig. 1.4b forms a parallelogram, and for this reason vectors are said to obey the parallelogram law of addition.

Problems

1.1   What is the weight of an object whose mass is 10 kg?

1.2   Ten seconds after starting from rest a car has a velocity of 30 m/s. What was its average acceleration?

1.3   The dial registers 180 lbf when a man stands on an accurate weighing machine. What is his (a) mass and (b) weight in SI units?

1.4   If an object of mass 50 kg were dropped down a mine shaft, what would be its initial acceleration?

1.5   A table is to be designed to support a mass of 100 kg. If the factor of safety is to be 1.5, what vertical force should the table be designed to withstand? (Note: The actual force should be multipled by the factor of safety to give the design force.)

1.6   Express 177×10^5 kg in megagrams; 0.000025 m in micrometres; 25 nm in metres.

1.7   Of what order of magnitude is the weight of an apple?

1.8   A little green man weighs 100 N on his home planet X. What will he weigh when he lands on Earth during an expedition in 2001? Note: The mass of planet X is 2.1×10^{24} kg, and its diameter is 3.2×10^6 m.

1.9 Referring to problem 1.8, assuming that the little green man's density is roughly equal to that of an Earth man and that he is roughly the same shape, what would be his height? (Assume a man 2 m tall weighs 900 N on Earth.)

2

Point Forces, Moments, and Static Equilibrium

2.1 Introduction

In this chapter, the methods of solution of problems where particles and bodies are in static equilibrium under the action of any system of coplanar forces will be described.

2.2 Force as a Vector

A force of magnitude 80 N applied at a point O in the direction $o \rightarrow a$ can be represented by the vector **oa** (Fig. 2.1). Similarly, a force of magnitude 40 N applied at the same point O in the direction $o \rightarrow b$ can be represented by the vector **ob**. If these two forces are added, then the result is the vector **oc** which, by trigonometry, has a length of $40\sqrt{3}$ N and acts in the direction $o \rightarrow c$. The latter force is the *resultant* of the two forces of 80 N and 40 N, and experiments have shown that the resultant, if applied alone, would have the same effect as the two original forces applied simultaneously.

It will be noted that in Fig. 2.1 a force acting at a point is represented by an arrow. Figure 2.2 shows some examples of the representation of forces. In these examples, the forces are assumed to act at specific points. However, these situations are idealized because, in fact, a force never acts at a point. For example, the wheel in Fig. 2.2a flattens where it contacts the road, and the force F_n is distributed over the small contact area. Similarly, the force F_w exerted by the mass on the cord in Fig. 2.2b is distributed over the cross-

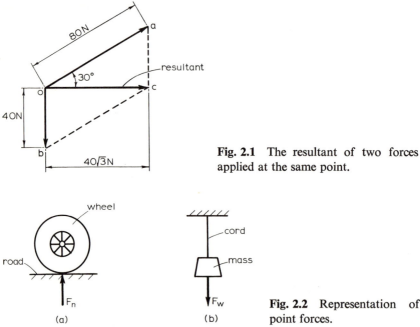

Fig. 2.1 The resultant of two forces applied at the same point.

Fig. 2.2 Representation of point forces.

sectional area of the cord. Distributed forces will be dealt with later, but in many practical problems in mechanics, the area over which a force is distributed is so small that it may be considered a point. Thus, we shall deal first with the theory of *point forces*.

2.3 Resultant of Concurrent Coplanar Forces

When a system of several forces acts through a single point, these forces are said to be concurrent. Also, when the directions in which all the forces act lie in one plane, these forces are said to be coplanar. Figure 2.3a shows two concurrent forces of magnitudes F_1 and F_2 acting at point O; the directions in which they act lie in the plane of the paper. The resultant of these two forces is the single force that may be used to replace them. The magnitude of this

Fig. 2.3 Concurrent, coplanar forces.

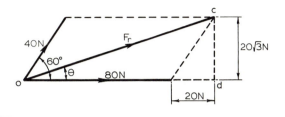

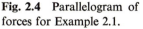

Fig. 2.4   Parallelogram of forces for Example 2.1.

resultant F_r is found by vector addition of the vectors $\mathbf{F}_1$ and $\mathbf{F}_2$ as shown in Fig. 2.3b; that is,

$$\mathbf{F}_r = \mathbf{F}_1 + \mathbf{F}_2 \tag{2.1}$$

Example 2.1   Forces with magnitudes 80 N and 40 N act at a point O. The angle between these forces is 60°. Find the magnitude and direction of their resultant.

Solution.   The parallelogram of forces is drawn in Fig. 2.4, and it can be seen that the magnitude of the resultant F_r is equal to the length of the hypotenuse of the right triangle *ocd*. Thus, the magnitude of the resultant is given by

$$F_r = \sqrt{(100)^2 + (20\sqrt{3})^2}$$
$$= 105.83 \text{ N}$$

Also, the direction of the resultant is given by

$$\theta = \arctan \frac{20\sqrt{3}}{100} = 19.11°$$

2.4   Resolution of Forces

The replacement of two or more forces by a single force having the same effect (resultant) is called the *composition* of forces. An inverse procedure is where a single force is replaced by two or more forces having the same effect. This is called the *resolution* of forces, and the latter forces are known as *components*. Usually, a force is replaced by two components acting at right angles to each other. Figure 2.5 shows a force of magnitude F which may be replaced by the components of magnitudes F_x and F_y along the x and y axes, respectively.

For the case shown, the components of F are

$$F_x = F \cos \theta \tag{2 2}$$

and
$$F_y = F \sin \theta \tag{2.3}$$

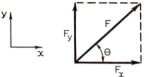

Fig. 2.5 Resolution of forces.

The resolved part of a force F in a given direction (the component) may be regarded as the effect of the force F in that direction.

2.5 Static Equilibrium of a Particle

A particle is a body of infinitesimally small size, and the forces acting on a particle are necessarily concurrent. If the resultant of several concurrent forces is zero, these forces will cancel each other completely and will produce no effect on the particle on which they act. The particle is then said to be in *static equilibrium*. Thus, a particle is in static equilibrium if the vector sum of the forces acting on it is zero. It follows from this that, for static equilibrium of a particle, the sum of the components of the forces in any given direction must be zero. Using the Greek letter Σ, meaning "summation," we can write, for static equilibrium,

$$\sum F_x = 0 \qquad\qquad (2.4)$$

$$\sum F_y = 0 \qquad\qquad (2.5)$$

where $\sum F_x$ is the sum of the components of all the forces in the x direction and $\sum F_y$ is the sum of the components of all the forces in the y direction.

A force which, when applied, produces static equilibrium of a body is called the *equilibrant*. The equilibrant is equal in magnitude but opposite in direction to the resultant of the remaining forces acting on the body. For example, in Fig. 2.6, F_1 and F_2 are two known forces acting on a particle. The resultant of these two forces is F_r, and the equilibrant F_e is equal and opposite to F_r.

Example 2.2   Figure 2.7 shows two concurrent forces having magnitudes of 80 N and 100 N. Find the magnitude of the equilibrant force F_e and the direction in which it acts.

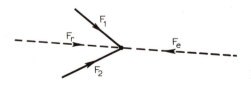

Fig. 2.6 Equilibrant of two forces.

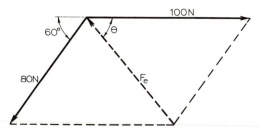

Fig. 2.7 Forces in Example 2.2.

Solution. Since the equilibrant must be equal and opposite to the resultant, it is directed as shown in Fig. 2.7. Now resolving components along the x and y axes gives

$$\sum F_x = 0: \qquad 100 - 80 \cos 60° - F_e \cos \theta = 0 \qquad \text{(a)}$$

$$\sum F_y = 0: \qquad -80 \sin 60° + F_e \sin \theta = 0 \qquad \text{(b)}$$

From Eq. (a)

$$F_e \cos \theta = 100 - 80(0.5) = 60.0 \qquad \text{(c)}$$

while from Eq. (b)

$$F_e \sin \theta = 80(0.866) = 69.3 \qquad \text{(d)}$$

Thus, dividing Eq. (d) by Eq. (c) gives

$$\tan \theta = \frac{69.3}{60.0} = 1.16$$

or $\qquad\qquad\qquad\qquad \theta = 49.1° \qquad\qquad\qquad\qquad\qquad \text{(e)}$

Substituting this value of θ into Eq. (d) and solving for F_e, we get

$$F_e = \frac{69.3}{\sin 49.1°} = 91.5 \text{ N}$$

Alternatively, since the vector sum of the two forces and the equilibrant F_e must be zero, the vector diagram of these three forces must close with the arrows following one another around the diagram. Thus, drawing a vector diagram to scale (Fig. 2.8) gives

$$F_e = 91.5 \text{ N}$$

and $\qquad\qquad\qquad\qquad \theta = 49.1°$

2.6 Free-Body Diagram

A mass is suspended from a ring which in turn is supported by two wires AC and BC as shown in Fig. 2.9. The forces in the wires can be found by con-

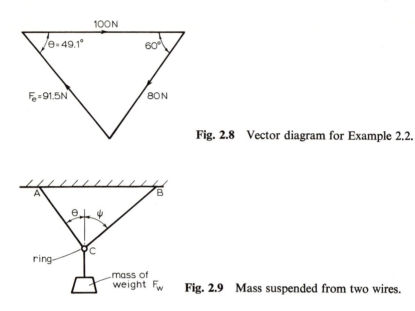

Fig. 2.8   Vector diagram for Example 2.2.

Fig. 2.9   Mass suspended from two wires.

sidering the static equilibrium of the ring to which the wires are connected. It is clear that wires *AC* and *BC* are pulling upward on the ring in the directions in which the wires lie and that the mass is pulling vertically downward on the ring. This situation can be illustrated by the *free-body diagram* in Fig. 2.10a where the ring, isolated from its surroundings, is assumed to be a particle and where all the forces acting on the ring are shown as labeled arrows. Figure 2.10b shows the corresponding *vector diagram* which could be used to obtain a graphical solution if the magnitudes of the angles were known.

For many problems in mechanics, the construction of the free-body diagram is an essential step in their correct solution. In fact, the most difficult step in many problems is translating a practical situation into an appropriate free-body diagram.

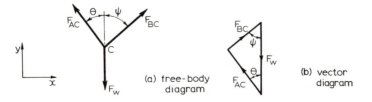

Fig. 2.10   Forces acting on the ring in Fig. 2.9.

Returning to the example, the solution (in general terms) can now be obtained by resolving forces along the x and y axes, respectively, and applying Eqs. (2.4) and (2.5). Thus,

$$\sum F_x = 0: \qquad -F_{AC} \sin \theta + F_{BC} \sin \psi = 0 \qquad (2.6)$$

$$\sum F_y = 0: \qquad F_{AC} \cos \theta + F_{BC} \cos \psi - F_w = 0 \qquad (2.7)$$

Substitution for F_{BC} from Eq. (2.6) in Eq. (2.7) and rearrangement give

$$F_{AC} = \frac{F_w \sin \psi}{\sin (\theta + \psi)} \qquad (2.8)$$

Similarly, substitution for F_{AC} from Eq. (2.6) in Eq. (2.7) and rearrangement give

$$F_{BC} = \frac{F_w \sin \theta}{\sin (\theta + \psi)} \qquad (2.9)$$

Alternatively, the solution for F_{BC} [Eq. (2.9)] could have been obtained from the solution for F_{AC} [Eq. (2.8)] by interposing the angles θ and ψ.

Example 2.3   A simple boom is shown in Fig. 2.11. The boom is hinged to the wall at A, and its free end is supported by the cable as shown. A 100-kg mass is suspended from the free end of the boom. Determine the tension force in the cable.

Solution.   First the weight of the mass is calculated. Thus,

$$F_w = mg = 100(9.81) = 981 \text{ N}$$

Then the free-body diagram is drawn for point B (Fig. 2.12a) where F_t rep-

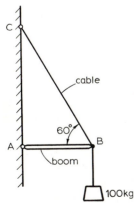

Fig. 2.11   Geometry for Example 2.3.

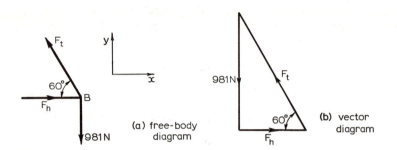

Fig. 2.12   Forces acting at point B in Fig. 2.11.

resents the tension force in the cable and F_h represents the horizontal force with which the end of the boom is pushing to the right.

Summing all the forces acting at B in the y direction and setting the result to zero, we get

$$\sum F_y = 0: \qquad F_t \sin 60° - 981 = 0$$

$$F_t = \frac{981}{\sin 60°} = \frac{981}{0.866} = 1133 \text{ N}$$

$$= 1.13 \text{ kN}$$

Alternatively, since point B is in static equilibrium, the vectors representing the three forces acting at this point must form a closed triangle, with the arrows following one another around the triangle. This is the vector diagram shown in Fig. 2.12b. The magnitude of the unknown force F_t may be found graphically by drawing the vector diagram to scale or by applying trigonometry to a sketch of the vector diagram. In the latter case we get

$$F_t = 981 \text{ cosec } 60°$$

$$= 1133 \text{ N}$$

2.7   Forces Acting on a Rigid Body

The straight line along which a force vector lies is called the *line of action* of the force. If two or more forces have a common line of action, they are said to be *collinear*.

Figures 2.13a and 2.13b show a rigid body with two forces acting on it. Such a rigid body is commonly called a two-force body. In the first case (Fig. 2.13a) the two forces are collinear but opposite in direction. It can be seen that the body will be in static equilibrium since there will be no tendency for it

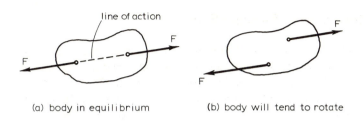

(a) body in equilibrium   (b) body will tend to rotate

Fig. 2.13 Two forces acting on a body.

to move. In the second case (Fig. 2.13b) the same forces are equal and opposite but not collinear (their lines of action do not coincide). Although the vector sum of these two forces is zero, the body is not in static equilibrium and will rotate in a clockwise direction. This rotation will occur because the two forces produce a turning effect or moment. Thus, if a two-force body is in static equilibrium, the two forces must have the same magnitude, have opposite directions, and be collinear.

2.8 Moment of a Force with Respect to an Axis

Figure 2.14 shows a single force F acting at point P on a rigid body. The perpendicular distance from its line of action to an axis z passing through O and perpendicular to the plane of the paper is r. This distance r is known as the *moment arm* of force F with respect to the axis z and is a nonnegative number. The *moment M* of force F with respect to (or about) axis z is defined by the equation

$$M = Fr \tag{2.10}$$

and in the example is a clockwise moment because the turning effect about axis z would be clockwise.

In two-dimensional problems (when the forces are coplanar) the axis z is adequately represented by the point O. Consequently, we would usually state that M, given by Eq. (2.10), is the moment of F about point O.

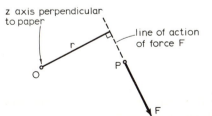

Fig. 2.14 Force having a moment about point O.

2.9 Static Equilibrium of a Rigid Body

A motionless body is in static equilibrium if it remains motionless. Clearly, if the combined effect of all the forces acting on a body produces a moment about any axis, the body will not be in static equilibrium. Thus, for static equilibrium of a rigid body, we have the additional condition that the sum of the moments of all the forces about any axis must be zero. Therefore,

$$\sum F_x = 0 \tag{2.11}$$

$$\sum F_y = 0 \tag{2.12}$$

and

$$\sum M = 0 \tag{2.13}$$

are the necessary conditions for static equilibrium of a rigid body under the action of coplanar forces.

In the solution of practical problems, the choice of the two axes x and y is purely arbitrary because if the sums of the force components in any two directions are zero, then the sum will be zero in any other direction. Similarly, if the sum of the moments of forces is zero about one axis, it will be zero about any other axis. Thus, the choice of the axis for considering moments is also arbitrary. In particular cases, therefore, the axes should be chosen so as to facilitate the solution of the problem.

Example 2.4 Find F_C, F_D, and θ for static equilibrium of the square plate shown in the free-body diagram in Fig. 2.15.

Solution. For static equilibrium of the plate, the vector sum of the forces and the sum of the moments acting on the body must be zero. Thus, resolving forces along the x and y axes, respectively, gives

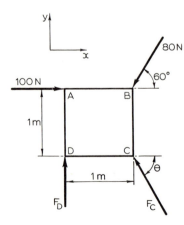

Fig. 2.15 Square plate with nonconcurrent coplanar forces acting.

$$\sum F_x = 0: \qquad 100 - 80 \cos 60° - F_C \cos \theta = 0 \qquad \text{(a)}$$
$$\sum F_y = 0: \qquad F_D + F_C \sin \theta - 80 \sin 60° = 0 \qquad \text{(b)}$$

Summing moments about C and setting the resultant moment to zero, we get

$$\sum M_C = 0: \qquad 100(1) + F_D(1) - 80 \cos 60°(1) = 0 \qquad \text{(c)}$$

From Eq. (c)

$$F_D = -100 + 40 = -60 \text{ N} \qquad \text{(d)}$$

Substitution of Eq. (d) into Eq. (b) gives

$$F_C \sin \theta = 69.3 + 60 = 129.3 \qquad \text{(e)}$$

From Eq. (a)

$$F_C \cos \theta = 100 - 40 = 60 \qquad \text{(f)}$$

Hence, dividing Eq. (e) by Eq. (f), we get

$$\frac{F_C \sin \theta}{F_C \cos \theta} = \tan \theta = \frac{129.3}{60}$$

or

$$\theta = 65° \qquad \text{(g)}$$

From Eq. (f)

$$F_C = \frac{60}{\cos \theta} = \frac{60}{0.422} = 142 \text{ N} \qquad \text{(h)}$$

It should be noted that, in this solution, the point C was chosen for taking moments so that the unknowns F_C and θ would not appear in Eq. (c). It should also be noted that the moment of the 80-N force was obtained by considering the moments of its components (F_x and F_y along the x and y axes, respectively) and realizing that the moment of the component F_y is zero. Thus, an alternative to the condition $\sum M = 0$ is

$$\sum (xF_y - yF_x) = 0 \qquad \text{(2.14)}$$

where x and y are the moment arms of the force components from the origin of the coordinate system. This is illustrated in Fig. 2.16, which shows a force F and its two components F_x and F_y. If counterclockwise moments are considered positive, then the moment of the force F about O is given by

$$M = -Fr \qquad \text{(2.15)}$$

Since

$$r = c \cos \theta \qquad \text{(2.16)}$$

and

$$c = y - x \tan \theta \qquad \text{(2.17)}$$

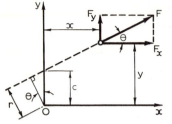

Fig. 2.16   Moment due to force components.

then
$$M = F(x \sin \theta - y \cos \theta) \tag{2.18}$$

or
$$M = xF \sin \theta - yF \cos \theta \tag{2.19}$$

Since

$$F \sin \theta = F_y, \qquad F \cos \theta = F_x$$

then
$$M = xF_y - yF_x \tag{2.20}$$

2.10   Couples

A couple is a pair of parallel forces which are equal in magnitude, opposite in sense, but noncollinear. Figure 2.17 shows two such forces acting on a body; the lines of action of the forces are a distance d apart. If we take the sum of the moments of these two forces about a point O, we get

$$M_O = F(d + x) - Fx$$
$$= Fd \qquad \text{(clockwise)} \tag{2.21}$$

Since the distance x from point A to point O does not appear in Eq. (2.21), the moment due to this couple is Fd about any axis or point. An alternative way of representing this couple is as shown in Fig. 2.18, where M is equal to Fd. It follows that, when considering the equilibrium of a body where a

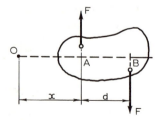

Fig. 2.17   Couple acting on a body.

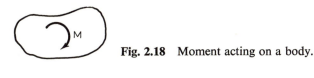

Fig. 2.18 Moment acting on a body.

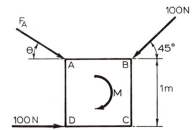

Fig. 2.19 Square plate with nonconcurrent coplanar forces and a moment acting.

couple or moment M is present, the moment M or the moment of the couple has the same magnitude regardless of the axis chosen for taking moments.

Example 2.5 Find the magnitude and sense of the moment M for equilibrium of the square plate shown in Fig. 2.19.

Solution. Taking moments about A and assuming clockwise moments are positive, we get

$$\sum M_A = 0: \qquad M + 1(100 \sin 45°) - 1(100) = 0$$

or $\qquad M = 100(1 - \sin 45°)$

$\qquad\qquad = 100(1 - 0.707) = 29.3 \text{ N·m} \qquad \text{(clockwise)}$

2.11 Special Cases of Coplanar Force Systems

2.11.1 Two Coplanar Forces

When a body is in equilibrium under the action of only two coplanar forces, the forces must be equal, collinear, and opposite in direction. It can be seen from Fig. 2.20 that if the forces F_1 and F_2 are not equal, the conditions that $\sum F_x$ and $\sum F_y$ equal zero would not be satisfied. Also it can be seen that if the lines of action of the two forces did not coincide, an unbalanced couple or moment would exist, and the condition that $\sum M$ equals zero would not be satisfied.

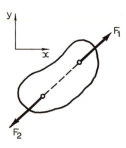

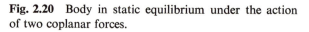

Fig. 2.20 Body in static equilibrium under the action of two coplanar forces.

2.11.2 Three Coplanar Forces

In Fig. 2.21, the lines of action of the forces F_1 and F_2 intersect at O, and thus their moments about O are zero. For equilibrium, the moment of the third force about O must be zero, and hence the line of action of the force F_3 must also pass through O. It can be concluded, therefore, that if a body is in equilibrium under the action of three coplanar forces, these forces must be concurrent; i.e., their lines of action must pass through the same point.

2.12 Resolution of a Force into a Force and a Couple

In the solution of some problems, it is useful to replace a single point force with a point force of the same magnitude acting at a different point but with a parallel line of action, together with a couple. Figure 2.22 illustrates this situation. Here, the force F acting at point P and having a moment arm r from the point O is replaced with a parallel force F acting at O and a couple of magnitude Fr.

Example 2.6   Figure 2.23 shows a bracket riveted to a vertical rolled steel stanchion. The effect of the 10-kN force can be considered as (a) a vertical

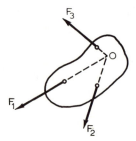

Fig. 2.21 Body in static equilibrium under the action of three coplanar forces.

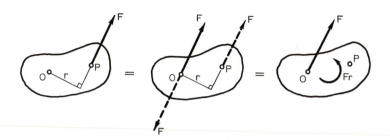

Fig. 2.22   Resolution of a force into a force and a couple.

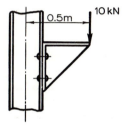

Fig. 2.23   Figure for Example 2.6.

force F acting at the centerline of the stanchion and providing compression and (b) a moment M providing bending. Find the values of F and M.

Solution.   The load systems for the original situation and the equivalent force F and moment M are shown in Fig. 2.24. Since the force systems in the two cases must be equivalent, they can be equated. Taking moments about O, we get

$$\sum M_O = 0: \qquad\qquad 10(0.5) = M$$

or $\qquad\qquad\qquad\qquad\qquad M = 5 \text{ kN} \cdot \text{m}$

Resolving forces vertically, we get

$$\sum F = 0: \qquad\qquad\qquad 10 = F$$

or $\qquad\qquad\qquad\qquad\qquad F = 10 \text{ kN}$

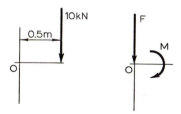

Fig. 2.24   Load systems for Example 2.6.

Problems

2.1 Forces of magnitudes 400 N and 200 N act at a point O (Fig. P2.1). The angle between these two forces is $75°$. Find the magnitudes and directions of their resultant and equilibrant forces.

2.2 Forces of magnitudes 500 N and 200 N act at a point O as shown in Fig. P2.2. Find the magnitudes and directions of their resultant and equilibrant forces.

2.3 Determine the resultant of the forces shown in Fig. P2.3. What are the magnitude and direction of the equilibrant force?

2.4 Referring to Fig. P2.4, resolve the force F into two components F_x and F_y.

2.5 What are (a) the magnitude and (b) the direction relative to the x axis of the resultant of the two forces shown in Fig. P2.5?

2.6 Find the magnitudes of the forces F_x and F_y so that the particle P (Fig. P2.6) will be in equilibrium.

2.7 The suspended object shown in Fig. P2.7 has a mass of 10 kg. What are the magnitudes of the tension forces F_{AC} and F_{BC} in the two suspension cords?

2.8 Referring to Fig. P2.8, at what angle θ must F_1 be applied in order to have the combined effect of F_1 and F_2 be equal to that of a 20-kN force?

2.9 Given the free-body diagram shown in Fig. P2.9, determine the unknown forces for static equilibrium.

2.10 Given the free-body diagram shown in Fig. P2.10, determine the unknown forces for static equilibrium.

Fig. P2.1   Fig. P2.2

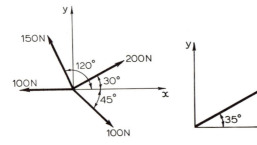

Fig. P2.3   Fig. P2.4

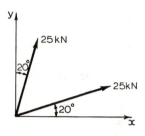

Fig. P2.5

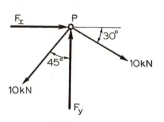

Fig. P2.6

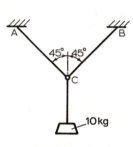

Fig. P2.7

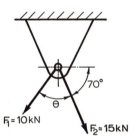

Fig. P2.8

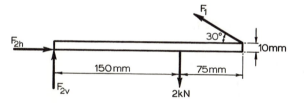

Fig. P2.9

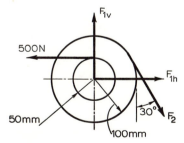

Fig. P2.10

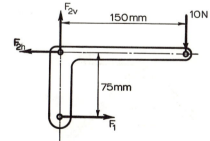

Fig. P2.11

2.11   Given the free-body diagram shown in Fig. P2.11, determine the unknown forces for static equilibrium.

2.12   Given the free-body diagram shown in Fig. P2.12, determine the unknown forces for static equilibrium.

2.13   Given the free-body diagram shown in Fig. P2.13, determine the unknown forces and moment for static equilibrium.

2.14   Given the free-body diagram shown in Fig. P2.14, find the forces F_C and F_B and the angle θ required to hold bar AB in static equilibrium.

2.15   Determine the magnitudes of the forces that the members AB and BC exert on the joint B in Fig. P2.15.

2.16   A weightless rope that is fastened at the points $(0, 0)$ and $(9, 0)$ carries a 100-N weight and an unknown weight F_w as shown in Fig. P2.16. The vertices of the rope polygon lie at the points $(7, 3)$ and $(4, 4)$. Determine the

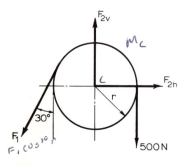

Fig. P2.12

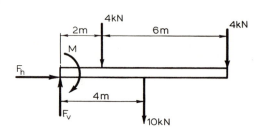

Fig. P2.13

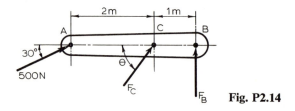

Fig. P2.14

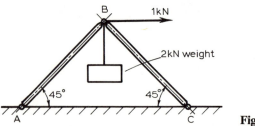

Fig. P2.15

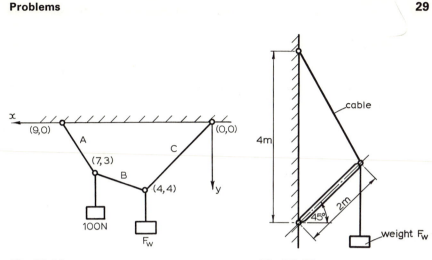

Fig. P2.16   Fig. P2.17

magnitudes of the forces F_A, F_B, and F_C in the three parts of the rope and the magnitude of the weight F_w.

2.17   Find the magnitude of the tension force F_t in the cable (Fig. P2.17) in terms of the weight of magnitude F_w.

2.18   The total length of the "light" cable (Fig. P2.18) is 60 m. Find the distance x for static equilibrium. [Notes: (1) The tension in the cable is the same on both sides of the pulley; (2) a "frictionless" pulley is one whose resistance to rolling is zero; (3) a "light" cable has zero mass.]

2.19   A light elastic string originally of length $2a$ is secured horizontally between two points without stretching. A weight F_w is suspended from a point midway

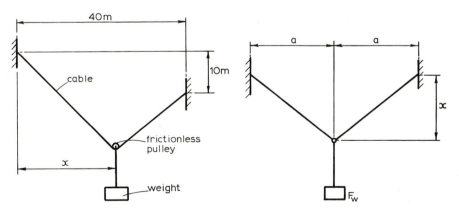

Fig. P2.18   Fig. P2.19

along the string as shown in Fig. P2.19. The extension of a portion of the string of length a under a tension F is given by akF. (a) Derive an expression for the deflection x of the center of the string in terms of a, F_w, and the elastic constant k for the string. (b) Calculate the deflection x when $a = 2$ m, $F_w = 10$ N, and $k = 0.1$ N^{-1}.

3

The Free-Body Diagram

3.1 Introduction

The methods of solution of problems where bodies are in static equilibrium under the action of any system of coplanar forces have now been described. However, it has already been mentioned that when presented with practical situations, the first, and possibly most difficult, step is to draw the appropriate free-body diagram. This diagram shows the body in the position under consideration isolated from all surrounding bodies and with all the forces acting on the body in the form of labeled arrows. These forces can arise in various ways, and some of the more common situations will be discussed in this chapter.

3.2 Forces at Supports

Figure 3.1 shows a pin-jointed truss. It consists of several straight members joined to each other by means of pin joints and supported at two points, *A* and *B*. A pin joint is effectively a hinge which provides no frictional resistance to rotation. Like many other concepts in mechanics a frictionless hinge is an idealization and cannot really exist. However, often the assumption of a frictionless hinge approaches the real situation so closely that the error involved is quite negligible. A true pin joint connecting two members might look like the illustration in Fig. 3.2. With such a joint and in the absence of friction no moment can be transmitted in a plane perpendicular to the pin.

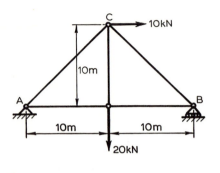

Fig. 3.1 Pin-jointed truss.

Fig. 3.2 Pin joint.

Returning to the drawing of the truss (Fig. 3.1), it can be seen that the corner A is pinned to a fixture and that an unknown direct force can be applied by the fixture to the frame at that point. This force (or reaction) F_A can be illustrated as shown in Fig. 3.3a. Alternatively, and usually more conveniently, this unknown force can be represented by its horizontal and vertical components F_{Ah} and F_{Av}, respectively, as shown in Fig. 3.3b, or by components in any two orthogonal directions, such as along the x and y axes, which may or may not coincide with horizontal and vertical directions, respectively. The corner B of the truss is supported differently; it can be seen that it is both pinned and resting on rollers which would present no resistance to motion of the truss in the horizontal direction at that point. This means there can be no horizontal component of the reaction at the support. In other words, the force exerted by the fixture on the truss must be vertical, as shown in Fig. 3.3c.

Now, by considering the whole truss ABC as a rigid body and neglecting its weight, the free-body diagram can be drawn as shown in Fig. 3.4. We are now in a position to solve for the unknown reactions using the techniques already described. Taking moments about A, we get

$$\sum M_A = 0: \qquad 20F_B - 10(10) - 10(20) = 0$$

$$F_B = 15 \text{ kN}$$

It should be noted that the point A was chosen to eliminate as many unknowns as possible.

Resolving forces horizontally, we get

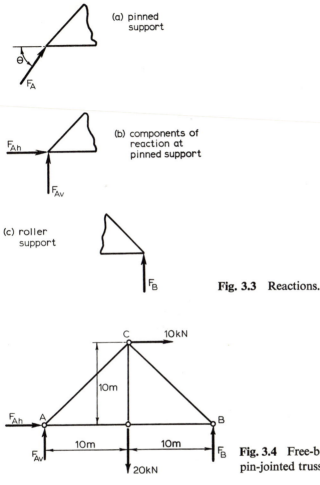

(a) pinned
   support

(b) components of
   reaction at
   pinned support

(c) roller
   support

Fig. 3.3 Reactions.

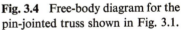

Fig. 3.4 Free-body diagram for the pin-jointed truss shown in Fig. 3.1.

$$\sum F_h = 0: \qquad\qquad F_{Ah} + 10 = 0$$
$$F_{Ah} = -10 \text{ kN}$$

Since this answer is negative, it simply means that the arrow showing F_{Ah} in the free-body diagram was drawn acting in the wrong direction; it actually acts to the left instead of the right.

Resolving forces vertically, we get

$$\sum F_v = 0: \qquad\qquad F_{Av} - 20 + F_B = 0$$
$$F_{Av} = 20 - F_B = 20 - 15 = 5 \text{ kN}$$

The vertical component of the reaction at B does indeed act upward.

3.3 Forces Applied Through Cables or Pin-Jointed Members

Figure 3.5 shows a truss ABC supported in a different way. The corner A is
pinned as before, but instead of a roller support at B, the corner C is connected to the wall (or fixture) at D by means of a pin-jointed member or cable.
If a free-body diagram is drawn for the member DC (Fig. 3.6a), it will be seen

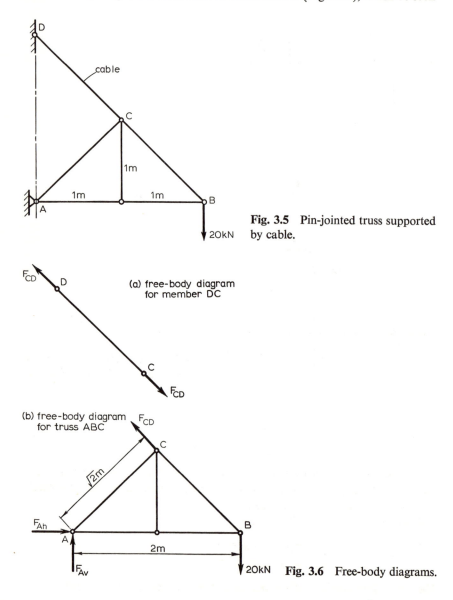

Fig. 3.5 Pin-jointed truss supported
by cable.

(a) free-body diagram
for member DC

(b) free-body diagram
for truss ABC

Fig. 3.6 Free-body diagrams.

that the forces acting on the member at D and C must be collinear. (It was shown in Chapter 2 that if only two forces act on a member, these forces must be collinear and opposite in sense.) This means that the line of action of the force exerted on the truss ABC at C must pass through D. Knowing this the free-body diagram for the truss ABC can now be drawn (Fig. 3.6b). Taking moments about A, we get

$$\sum M_A = 0: \qquad \sqrt{2}F_{CD} - 20(2) = 0$$

$$F_{CD} = \frac{20(2)}{\sqrt{2}} = 28.3 \text{ kN}$$

Resolving horizontally, we get

$$\sum F_h = 0: \qquad F_{Ah} - \frac{F_{CD}}{\sqrt{2}} = 0$$

$$F_{Ah} = \frac{F_{CD}}{\sqrt{2}} = \frac{20(2)}{\sqrt{2}(\sqrt{2})} = 20 \text{ kN}$$

Resolving vertically, we get

$$\sum F_v = 0: \qquad F_{Av} - 20 + \frac{F_{CD}}{\sqrt{2}} = 0$$

$$F_{Av} = 20 - 20 = 0$$

It is interesting to note that the fact that F_{Av} is zero could have been deduced with the realization that the truss ABC has only three coplanar forces acting on it. This means that the line of action of the reaction at A must pass through point B, which is the intersection of the lines of action of the other two forces; therefore in this example the reaction at A must be horizontal.

3.4   Forces Due to Contact with Other Bodies

Figure 3.7 shows two bodies A and B in contact at point P. The tangent to the two surfaces at the contact point is also shown. In general, body B will be pushing body A with a force F. However, it is usually more convenient to employ the components of F tangential to and normal to the contact, i.e., the tangential force F_f and the normal force F_n, respectively. The tangential force F_f arises only because of the resistance to sliding motion between the two bodies. Resistance to sliding is called friction, and for this reason the force F_f is often referred to as the frictional force or the frictional component.

It is important to realize that the frictional force on a body always opposes the motion of that body or the tendency for motion relative to the body in contact. Referring again to Fig. 3.7, if body B is pushing body A to the left,

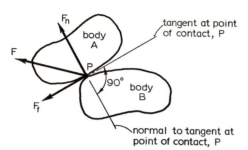

Fig. 3.7 Forces acting on body *A* due to contact with body *B*.

then body *A* will be tending to move to the right relative to body *B*. Thus, the frictional force on body *A* will be acting in the opposite direction (to the left) as shown. In practice, there is always a limit to the frictional force that a contact can withstand; this limiting frictional condition will be dealt with in Chapter 4.

In some problems it is useful to consider idealized frictionless surfaces. This means that the contact is not capable of resisting any tendency for relative sliding motion and that the frictional component F_f of the reaction will always be zero. One example of this was the roller support described earlier. Further examples are given in Fig. 3.8.

Example 3.1   Find the magnitude of the maximum force F_p that can be applied to the handle of the wrench (Fig. 3.9a) if the contact force between the nut and the wrench jaws is limited to 7.5 kN. Assume that the contact surfaces are frictionless.

Solution.   It is clear that as the force F_p is applied contact between the wrench jaws and nut will occur at the corners *A* and *B*. Also, since the surfaces are frictionless, the reactions at *A* and *B* will be perpendicular to the surfaces of the jaws. Thus, the free-body diagram for the wrench is as shown in Fig. 3.9b.

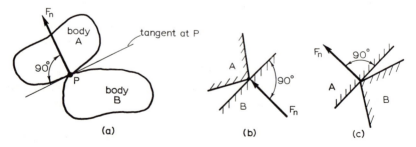

Fig. 3.8   Forces acting on body *A* when surfaces are frictionless.

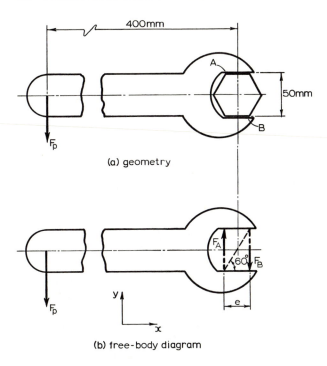

(a) geometry

(b) free-body diagram

Fig. 3.9 Wrench and nut.

This is a special case of a body with only three forces acting. Here the forces are parallel and, therefore, intersect at infinity. In this case only two equilibrium equations can be written; i.e., $\sum F_y$ and $\sum M$ are equal to zero. Summing the forces in the y direction gives

$$\sum F_y = 0: \qquad F_A - F_p - F_B = 0$$

$$F_A = F_p + F_B \qquad\qquad \text{(a)}$$

Taking moments about A, we get

$$\sum M_A = 0: \qquad F_B e - F_p\left(400 - \frac{e}{2}\right) = 0$$

$$F_B e = F_p\left(400 - \frac{e}{2}\right) \qquad\qquad \text{(b)}$$

From Eq. (a) it can be seen that F_A will always be the larger of the two reactions, and, thus, setting F_A equal to 7.5 kN will give the required value for F_p. Substitution of Eq. (b) into Eq. (a) gives

$$F_A = F_p + \frac{F_p}{e}\left(400 - \frac{e}{2}\right)$$

or

$$F_p = \frac{eF_A}{400 + (e/2)} \tag{c}$$

Since F_A is equal to 7.5 kN and e is equal to $50/\sqrt{3}$,

$$F_p = \frac{50}{\sqrt{3}}\left[\frac{7.5}{400 + (25/\sqrt{3})}\right] = 0.52 \text{ kN}$$

3.5   Forces Due to Gravity

In all the preceding examples we have neglected the weight of the body under consideration. In many situations this is acceptable because the external forces acting on a body are much larger than its weight. In all other cases the weight of a body must be shown on the free-body diagram and included in the analysis. For this purpose it is necessary to know the position of the *center of gravity* of the body. The center of gravity is defined as the point through which the line of action of the weight of the body may be assumed to act. For bodies of uniform density the center of gravity is at the geometric center (centroid) of the body. Determination of the position of the center of gravity for other than simple shapes requires special techniques and will be dealt with in Chapter 5.

Example 3.2   Draw free-body diagrams for the arrangement of two 50-kg cylindrical rollers shown in Fig. 3.10 and determine the magnitudes of all of the contact forces. Assume all surfaces are frictionless.

Solution.   Figures 3.11a and 3.11b show free-body diagrams for each roller separately and for the two rollers together, respectively. For the two rollers together (Fig. 3.11b) summing forces in the vertical and horizontal directions

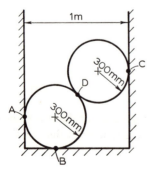

Fig. 3.10   Geometry for Example 3.2.

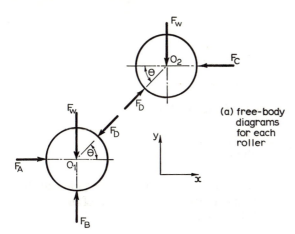

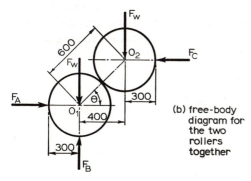

(a) free-body
diagrams
for each
roller

(b) free-body
diagram for
the two
rollers
together

Fig. 3.11   Free-body diagrams for Example 3.2.

gives

$\sum F_y = 0$: $F_B = 2F_w = 2(50)(9.81) = 981$ N   (a)

$\sum F_x = 0$: $F_A = F_C$ (b)

Taking moments about O_1 gives

$\sum M_{O_1} = 0$: $F_C\sqrt{(600)^2 - (400)^2} - F_w(400) = 0$

$$F_C = \frac{50(9.81)(400)}{447.2} = 439 \text{ N} \qquad (c)$$

Hence, from Eqs. (b) and (c),

$$F_A = 439 \text{ N}$$

For the uppermost roller (Fig. 3.11a), resolving forces in the vertical direction gives

$\sum F_y = 0$: $F_D \sin \theta = 50(9.81) = 490.5$ (d)

From Fig. 3.11b it is seen that

$$\cos \theta = \frac{400}{600}$$

$$\theta = 48.2°$$

Finally, substitution of this value for θ into Eq. (d) gives

$$F_D = \frac{490.5}{\sin 48.2°} = 658 \text{ N}$$

Example 3.3   The slender plank shown in Fig. 3.12a is held in equilibrium by the action of the force F_p. Find the magnitude of F_p if the mass of the plank is 200 kg. Assume all surfaces are frictionless.

Solution.   Clearly, the center of gravity of the plank lies halfway along its length, the reaction at B is perpendicular to the plank, and the reaction at A is perpendicular to the floor. Thus, the free-body diagram for the plank is as shown in Fig. 3.12b. Resolving horizontally, we get

$$\sum F_x = 0: \qquad\qquad F_p - F_B \sin \theta = 0$$

$$F_p = F_B \sin \theta \qquad\qquad\qquad (a)$$

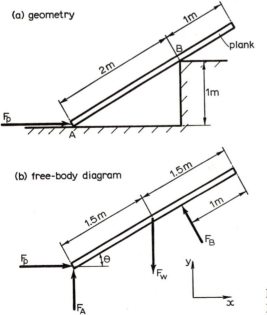

(a) geometry

1m

plank

2m

B

1m

F_p

A

(b) free-body diagram

1.5m

1.5m

1m

F_B

F_p

θ

F_w

y

F_A

x

Fig. 3.12   Diagrams for Example 3.3.

Taking moments about A, we get

$$\sum M_A = 0: \qquad\qquad F_w \cos \theta (1.5) - F_B(2) = 0$$

or
$$F_B = \frac{1.5}{2} F_w \cos \theta \qquad\qquad\text{(b)}$$

Substitution of Eq. (b) into Eq. (a) gives

$$F_p = \frac{1.5}{2} F_w \sin \theta \cos \theta \qquad\qquad\text{(c)}$$

Since

$$F_w = mg = 200(9.81) = 1962 \text{ N}$$

and, by inspection,

$$\theta = 30°$$

we get

$$F_p = \frac{1.5}{2}(1962) \sin 30° \cos 30°$$

$$= 637 \text{ N}$$

Alternatively, this problem could have been solved graphically. Since there are only three forces acting on the plank, these forces must be concurrent. Thus, the vector diagram is as shown in Fig. 3.13.

3.6   Forces Due to Built-in or Rigid Supports

Figure 3.14 shows a pole or beam held rigidly in the ground. The ground or rigid support can exert both a direct force (components F_h and F_v) and a couple M, to prevent rotation, on the lower end of the beam as shown.

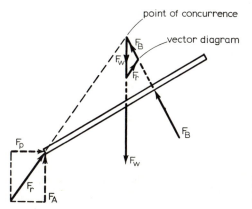

Fig. 3.13 Graphical solution to Example 3.3.

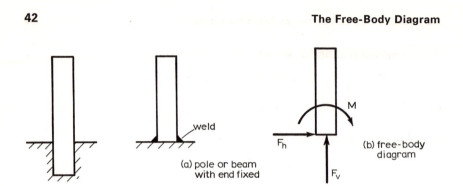

Fig. 3.14 Built-in support.

Example 3.4   Obtain the magnitudes of the horizontal (F_{Ah}) and vertical (F_{Av}) reactions and the fixing moment (M) for the end of the cantilever beam shown in Fig. 3.15a.

Solution.   The beam is imagined isolated from the wall, and the free-body diagram is drawn as shown in Fig. 3.15b. Resolving horizontally, we get

$$\sum F_x = 0: \qquad\qquad F_{Ah} - 10 \cos 45° = 0$$

$$F_{Ah} = 10 \cos 45° = 7.07 \text{ kN}$$

Resolving vertically, we get

$$\sum F_y = 0: \qquad\qquad F_{Av} - 10 \sin 45° - 10 = 0$$

$$F_{Av} = 10 \sin 45° + 10 = 17.07 \text{ kN}$$

(a) geometry

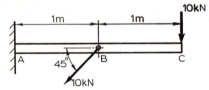

(b) free-body diagram

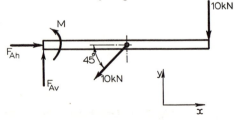

Fig. 3.15   Cantilever beam.

Taking moments about A, we get

$$\sum M_A = 0: \quad M - 10 \sin 45°(1) - 10(2) = 0$$

$$M = 10 \sin 45° + 20 = 27.07 \text{ kN} \cdot \text{m}$$

$$\text{(counterclockwise)}$$

Problems

3.1 Draw a free-body diagram for the homogeneous sphere of mass m resting against the frictionless surfaces as shown in Fig. P3.1.

3.2 Draw a free-body diagram for the uniform rod of mass m resting gainst the smooth surfaces as shown in Fig. P3.2. Note: Smooth means frictionless.

3.3 Draw a free-body diagram for the member AB shown in Fig. P3.3. Neglect the weight of the member. (Show only one force at point B.)

3.4 Draw a free-body diagram for a homogeneous cylindrical body of mass m resting against a step on an inclined plane as shown in Fig. P3.4. Neglect friction.

3.5 Draw a free-body diagram for the smooth uniform rod resting inside the

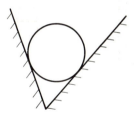

Fig. P3.1

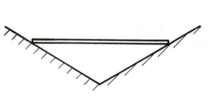

Fig. P3.2

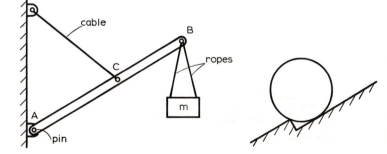

Fig. P3.3　　　　　　　　　**Fig. P3.4**

smooth hemispherical bowl shown in Fig. P3.5. Note: Smooth means frictionless.

3.6 Draw a free-body diagram for the arm of the balance scale shown in Fig. P3.6. Neglect the weight of the arm, and show only one force at $\mathscr{B}$.

3.7 Draw a free-body diagram for the homogeneous sphere of mass m resting against the frictionless wall shown in Fig. P3.7.

3.8 Draw a free-body diagram for the wheelbarrow shown in Fig. P3.8.

3.9 The 3-metre-long truck bed of a dump truck is held in equilibrium in the position shown in Fig. P3.9. If the truck bed and load weigh 5 kN, draw the free-body diagram for the truck bed, and determine the magnitude of the reaction at A and the force in the hydraulic actuator.

3.10 A force of 500 N is applied to the bar shown in Fig. P3.10. Draw a free-body diagram for the bar, and determine the force in the cable if the weight of the bar is negligible.

3.11 Figure P3.11 shows a bar AB and a cable used to support a barrel whose total

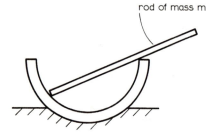

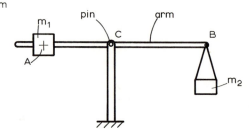

Fig. P3.5   **Fig. P3.6**

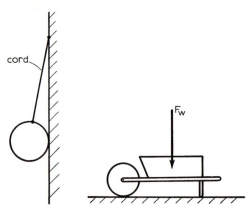

Fig. P3.7   **Fig. P3.8**

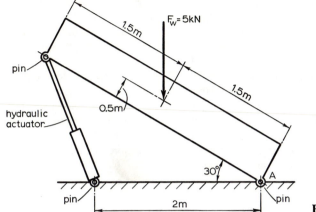

Fig. P3.9

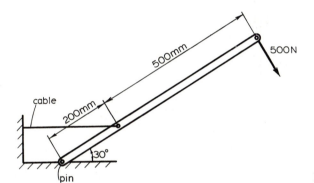

Fig. P3.10

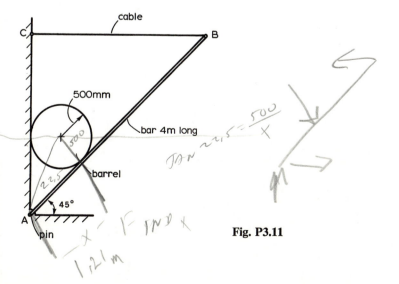

Fig. P3.11

weight, including its contents, is 2 kN. Draw free-body diagrams for the barrel and the bar, and determine the tension in the cable and the magnitude of the reaction at A. Neglect friction and the weights of the cable and bar.

3.12 For the two brackets shown in Fig. P3.12, determine the magnitudes of the reactions at O.

3.13 For the ratchet and pawl shown in Fig. P3.13, draw appropriate free-body diagrams, and hence determine the magnitude of the reaction at O.

3.14 Figure P3.14 shows a bracket, pin-supported at A and with a roller at B. If a load of 16 kN is applied at C, draw a free-body diagram for the bracket, and determine the magnitudes of the reactions at A and B.

3.15 What force F_p is needed to keep the mass in the position shown in Fig. P3.15? Also determine the tensions in cables AC and BC.

(a)   (b)

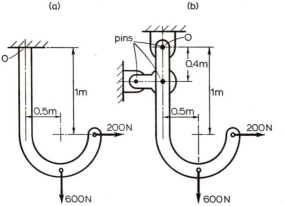

Fig. P3.12

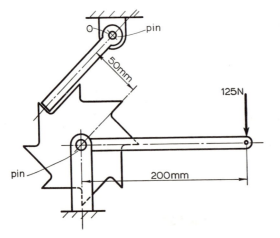

Fig. P3.13

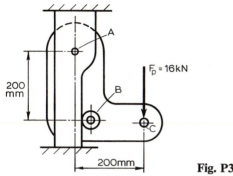

Fig. P3.14

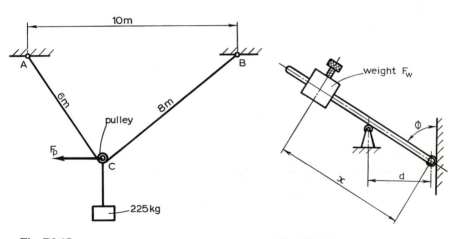

Fig. P3.15

Fig. P3.16

3.16 Referring to Fig. P3.16, neglecting the diameters of the rollers and the weights of the bar and roller, obtain an expression for x to give equilibrium. Neglect friction.

3.17 The thin plate shown in Fig. P3.17 weighs 500 N. If all surfaces are smooth, what are the magnitudes of the reactions at A, B, and C?

3.18 The beam shown in Fig. P3.18 has a mass of 200 kg. Find the magnitudes of the reactions at A and B.

3.19 The three-hinged circular arch shown in Fig. P3.19 carries a load of 10 kN. Find the magnitudes of the vertical and horizontal components of the reactions at A and C.

3.20 The weight of the coal truck shown in Fig. P3.20 is 20 kN. What force F_p should be applied to the cable to hold the truck stationary on the inclined track? Neglect friction.

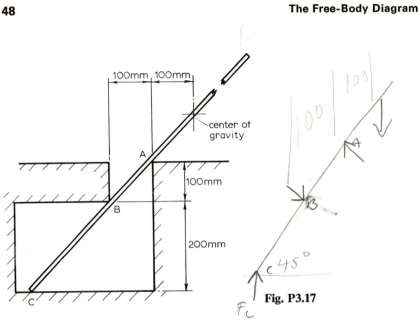

Fig. P3.17

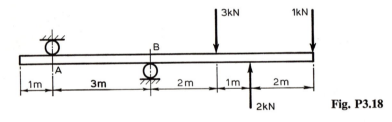

Fig. P3.18

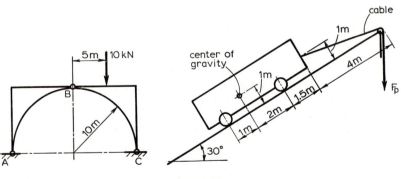

Fig. P3.19   Fig. P3.20

3.21 The 50-kg uniform ladder shown in Fig. P3.21 rests against a vertical wall at *A* and on the horizontal at *B*. The cord prevents the ladder from slipping. Determine the magnitudes of the reactions at *A* and *B* and the tension in the cord. Assume that all surfaces are frictionless.

3.22 A target at a fairground is arranged as shown in Fig. P3.22. It is to be designed such that if a missile hits the target with sufficient force to rotate the vertical support post through an angle of 45°, the target will fall to the ground. The extension of the elastic cord under load is given by lkF_t, where l (m) is its original length, F_t (N) is the tension in the cord, and k (1/N) is the elastic constant. When the target is upright the tension in the cord is zero. What should be the weight of the target F_w if $k = 0.01 \text{ N}^{-1}$? Neglect the weight of the support post.

3.23 Figure P3.23 shows a typical tone arm arrangement for a modern record player. The stylus "tracking" force, i.e., the vertical force between the stylus

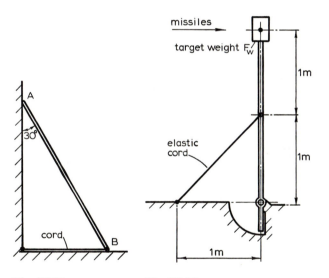

Fig. P3.21   **Fig. P3.22**

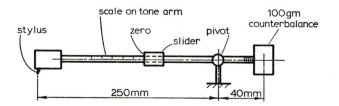

Fig. P3.23

and the record, has to be adjusted to suit the particular stylus employed. Adjustment of the tracking force is accomplished by moving a slider along the tone arm, the latter being marked with a scale for tracking force starting at zero. The procedure in setting the correct tracking force is to first set the slider on zero and balance the complete tone arm about the pivot by screwing the counterbalance along the thread on the tone arm extension. Then the slider is moved to the appropriate setting on the scale. For the tone arm shown, determine the mass of the slider so that the scale for tracking force marked on the arm is 10 mN/20 mm. Neglect the horizontal force due to friction acting on the stylus when a record is played.

3.24 Derive expressions for the reactions at *A* and *B* for the arrangement shown in Fig. P3.24. Assume the masses of the beam and cable are negligible.

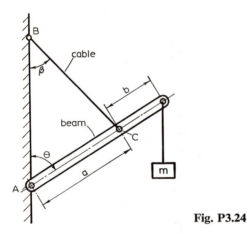

Fig. P3.24

4

Friction

4.1 Introduction

When a body slides on a surface the minute uneven and interlocking surface asperities on the contacting surfaces (Fig. 4.1) oppose the sliding motion. Some of the asperities weld together and then shear. This shearing provides the resistance to sliding and is represented by the horizontal components of the contact forces F_{r1}, F_{r2}, etc., which are referred to as the frictional forces or, more briefly, friction. Even so-called smooth surfaces have these microscopic asperities which cause friction, and this frictional resistance must be overcome during sliding.

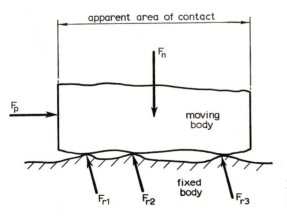

Fig. 4.1 Contacting asperities on sliding surfaces.

Friction is usually considered wasteful, as in automobiles, trains, and machines where it absorbs power and causes wear; in these circumstances efforts are made to reduce it. However, it can also be useful, such as in brakes, belt drives, and clutches. In fact, without friction one could not even walk or ride a bike because there would not be the necessary interaction between the floor and the shoe or the road and the tire. Without friction a bicycle wheel would just spin and skid; without friction a simple knot could not grip, a nail would be squeezed out of wood, and a nut would not stay on the end of a bolt.

4.2 Amontons' Law

According to Amontons' laws of dry (unlubricated) friction,

1. The frictional resistance F_f is proportional to the normal force F_n between the body and the surface on which it slides.
2. The frictional force is independent of the (apparent) area of contact.

Referring again to Fig. 4.1, it can be seen that contact between the two bodies occurs only at the tips of the asperities. This means that the real area of contact between the two bodies is much smaller than the apparent area of contact. If the normal force F_n is increased, the asperities deform, and the real area of contact increases. Thus, the total frictional force F_f required to shear these welded asperities also increases. From Amontons' first law this increase in frictional force F_f is proportional to the increase in normal force F_n.

Also, from this simplified explanation, it can be seen that the relationship between F_f and F_n is independent of the apparent area of contact. This is in accordance with Amontons' second law.

Figure 4.2 shows a free-body diagram for an object being pushed at constant speed across a horizontal rough surface. Resolving forces horizontally, we find that the force F_p required to push the object is equal and opposite to the frictional resistance F_f. From Amontons' law,

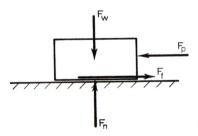

Fig. 4.2 Forces acting on sliding body.

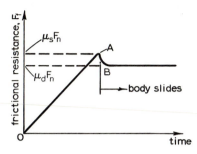

Fig. 4.3   Buildup of frictional resistance as body is pushed with gradually increasing force.

$$F_f \propto F_n$$

or
$$F_f = \mu_d F_n \qquad (4.1)$$

where μ_d is a constant and is known as the coefficient of dynamic friction between the two surfaces.

It should be noted that the frictional force F_f always opposes the motion of the body and that Eq. (4.1) applies only when the body is actually sliding. The graph in Fig. 4.3 shows what would happen if the force pushing the body were gradually increased from zero. Initially the body remains at rest since the force applied will be less than that required to overcome the maximum frictional resistance. Eventually, however, a point A on the graph is reached where the value of F_f has become equal to $\mu_s F_n$ (where μ_s is called the coefficient of static friction) and the body starts to slide. When the body starts to slide, the frictional resistance falls slightly and then becomes constant and equal to $\mu_d F_n$ [see Eq. (4.1)].

Thus, the limiting or maximum values of F_f are

during sliding: $\qquad\qquad F_f = \mu_d F_n$

for onset of sliding: $\qquad F_f = \mu_s F_n \qquad (4.2)$

Figure 4.4 shows the resultant F_r of the normal force F_n acting on the object and the force F_f resisting the constant speed motion of the object. The angle

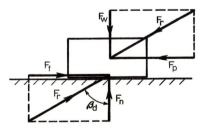

Fig. 4.4   Angle of dynamic friction for a body sliding over a surface.

β_d between the resultant F_r and a line normal to the surface is defined as the angle of dynamic friction. Now

$$\tan \beta_d = \frac{F_f}{F_n} = \frac{F_p}{F_w} = \mu_d$$

or

$$\beta_d = \arctan \mu_d \tag{4.3}$$

Similarly,

$$\beta_s = \arctan \mu_s \tag{4.4}$$

where β_s is the angle of static friction.

Example 4.1 What horizontal force is required to push a mass of 10 kg at constant speed across a horizontal surface if the coefficient of dynamic friction is 0.2?

Solution. From Fig. 4.5 it is seen that in this case F_n is equal to the weight mg of the mass. Thus, resolving horizontally, we get

$$F_p = F_f = \mu_d F_n = \mu_d mg$$
$$= 0.2(10)(9.81)$$
$$= 19.6 \text{ N}$$

Example 4.2 At what angle θ must a surface be tilted to just cause an object resting on it to slide if the coefficient of static friction is 0.3?

Solution. Figure 4.6 shows the free-body diagram for the object. Resolving forces along the plane gives

$$\sum F_x = 0: \qquad\qquad F_f = \mu_s F_n = mg \sin \theta \tag{a}$$

Resolving forces normal to the plane gives

$$\sum F_y = 0: \qquad\qquad F_n = mg \cos \theta \tag{b}$$

Substitution of Eq. (b) into Eq. (a) gives

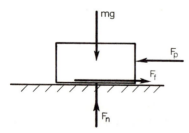

Fig. 4.5 Free-body diagram for Example 4.1.

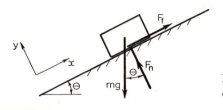

Fig. 4.6 Free-body diagram for Example 4.2.

$$\mu_s mg \cos \theta = mg \sin \theta$$

$$\theta = \arctan \mu_s = \arctan 0.3$$

$$\theta = 16.7°$$

In general, θ is equal to the angle of static friction β_s.

Example 4.3 A block is placed on a rough horizontal surface, and a gradually increasing horizontal force F_p is applied to the block. Under what conditions will the block slide or topple?

Solution. If the block is just about to slide, the normal reaction F_n and the frictional force F_f will both be distributed over its bottom surface (Fig. 4.7a). However, if the block is on the verge of toppling, these force components will both be concentrated at corner A (Fig. 4.7b). Referring to Fig. 4.7a, we get

$$\sum F_x = 0: \qquad\qquad F_p = F_f \qquad\qquad \text{(a)}$$
$$\sum F_y = 0: \qquad\qquad F_n = F_w \qquad\qquad \text{(b)}$$

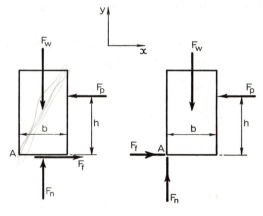

(a) block just about to slide

(b) block just about to topple

Fig. 4.7 Free-body diagrams for Example 4.3.

Since sliding is about to occur, we can write

$$F_f = \mu_s F_n \tag{c}$$

and combining Eqs. (a), (b), and (c), we get

$$F_p = \mu_s F_w \tag{d}$$

For any magnitude of F_p less than that given by Eq. (d) sliding cannot occur. However, depending on the geometry and the value of μ_s, the block may topple before it slides. Referring now to Fig. 4.7b and taking moments about A, we get

$$\sum M_A = 0: \qquad F_p h = \frac{F_w b}{2}$$

$$F_p = \frac{F_w}{2} \frac{b}{h} \tag{e}$$

for the block to topple. Thus, if the magnitude of F_p is less than that given by Eq. (e), the block cannot topple.

Comparing Eqs. (d) and (e), we can see that if h is large, the tendency will be for the block to topple before it slides. To find this critical value of h, we can combine Eqs. (d) and (e) and get

$$h = \frac{b}{2\mu_s} \tag{f}$$

Thus, if h is larger than the magnitude given by Eq. (f), the block will topple.

Example 4.4 Determine the maximum value of the distance d at which the lower end of the prop of negligible mass (Fig. 4.8a) can be set and still support the heavy hinged plank without slipping. The coefficient of static friction is 0.4.

Solution. The free-body diagram for the prop is shown in Fig. 4.8b. For slipping, F_f is equal to $\mu_s F_n$. Taking moments about A gives

$$\sum M_A = 0: \qquad F_f(900) \cos \theta - F_n(900) \sin \theta = 0$$

$$\tan \theta = \frac{F_f}{F_n} = \mu_s = 0.4$$

$$\theta = 21.8°$$

Now, by geometry,

$$\frac{d}{2} = 900 \sin \theta$$

Thus,

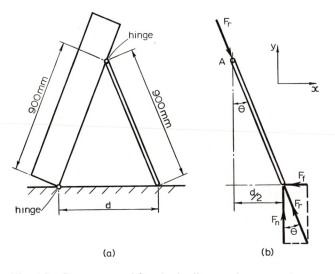

(a)   (b)

Fig. 4.8 Geometry and free-body diagram for Example 4.4.

$$d = 1800 \sin 21.8°$$

$$= 1800(0.371) = 669 \text{ mm}$$

Alternatively, since the prop has only two forces acting on it, these forces must be collinear. Hence, the reaction F_r must act at an angle θ to the vertical. Further, since the lower end of the prop is about to slide, this angle must also be equal to β_s. Hence,

$$\theta = \beta_s = \arctan \mu_s = 0.4$$

Example 4.5 If for the block on a rough inclined plane shown in Fig. 4.9 the values of the static and dynamic coefficients of friction are 0.3 and 0.2, respectively, find (a) the magnitude of the frictional force F_f acting on the block when F_p is 200 N and the block is at rest, (b) the magnitude of the force F_p to initiate motion of the block, and (c) the magnitude of the frictional force F_f when F_p is 600 N.

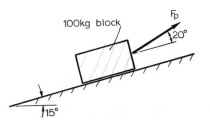

Fig. 4.9 Figure for Example 4.5.

Solution. (a) Figure 4.10a shows the free-body diagram. Resolving forces parallel to the plane, we get

$\sum F_x = 0$: $200 \cos 20° - F_f - mg \sin 15° = 0$

$200 \cos 20° - F_f - 100(9.81) \sin 15° = 0$

or $F_f = 200 \cos 20° - 981 \sin 15°$

$= -66 \text{ N}$ (i.e., up the plane)

(b) When motion is initiated (Fig. 4.10b) we can write

$$F_f = \mu_s F_n = 0.3 F_n \tag{a}$$

Thus, resolving parallel to the plane, we get

$\sum F_x = 0$: $F_p \cos 20° - F_f - 981 \sin 15° = 0$ \hfill (b)

Resolving perpendicular to the plane, we get

$\sum F_y = 0$: $F_p \sin 20° + F_n - 981 \cos 15° = 0$ \hfill (c)

Substitution of F_f from Eq. (a) into Eq. (b) gives

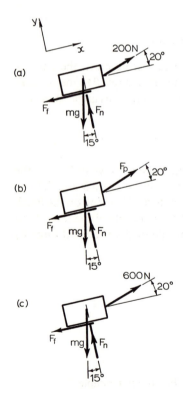

Fig. 4.10   Free-body diagrams for Example 4.5.

$$F_p \cos 20° = 0.3F_n + 981 \sin 15° \qquad \text{(d)}$$

Substitution of F_n from Eq. (c) gives

$$F_p \cos 20° = 0.3(981 \cos 15° - F_p \sin 20°) + 981 \sin 15°$$
$$0.940F_p = 284 - 0.103F_p + 254$$
$$1.043F_p = 538$$

Thus,

$$F_p = 516 \text{ N}$$

(c) When the force F_p is larger than 516 N the object will slide. Thus, referring to Fig. 4.10c, we can write

$$F_f = \mu_d F_n = 0.2F_n \qquad \text{(e)}$$

Resolving perpendicular to the plane, we get

$$\sum F_y = 0: \qquad 600 \sin 20° + F_n - 981 \cos 15° = 0$$

or
$$F_n = 981 \cos 15° - 600 \sin 20°$$
$$= 742 \text{ N}$$

Hence, substitution of this value into Eq. (e) gives

$$F_f = 0.2F_n = 0.2(742)$$
$$= 148 \text{ N}$$

4.3 Rolling Resistance

We know from experience that it is generally easier to roll a wheel across a surface than to slide a block across the same surface. Whereas the block encounters considerable resistance due to friction, the wheel rolls over the surface with little relative sliding motion at the contacting surfaces. However, even the wheel encounters some resistance due to the deformation of the wheel and the surface. For example, if a hard wheel moves over the surface of a soft material, the wheel deforms the material ahead of it, causing a reaction force F_r as shown in Fig. 4.11. This reaction force has a horizontal component F_f which causes resistance to rolling and can be considered equivalent to the frictional force arising in sliding friction, although it will generally have a much smaller magnitude. A similar effect occurs with a soft wheel rolling on a hard surface. In this case, however, the wheel deforms instead of the surface. Thus, the harder the materials of which the wheel and surface are made, the smaller the rolling resistance. Hence, with a steel railroad wheel on a steel track the rolling resistance is relatively small.

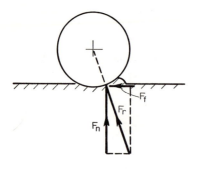

Fig. 4.11 Resistance due to rolling.

Problems

4.1 Determine the mass of the weight C to just turn the wheel B if the coefficient of static friction between the bar AD and the wheel is 0.4 (Fig. P4.1). Neglect the friction in the wheel bearing, and neglect the weights of the bar and cable.

4.2 Figure P4.2 shows an automobile on a hill. If the coefficient of static friction betwen the tires and the road is 0.7, determine the inclination θ of the steepest hill that the car can climb using (a) rear-wheel drive, (b) front-wheel drive, and (c) four-wheel drive.

4.3 Figure P4.3 shows a device that is to be used to secure a bar. For the loading shown, determine the reaction at A if the coefficient of static friction between the bar and the cams is 0.4.

4.4 Figure P4.4 shows a movable bracket which is used to support a load F_w. If the coefficient of static friction between the bracket and the support column is 0.3, what is the minimum value of a to prevent sliding of the bracket down the column? Neglect the weight of the bracket.

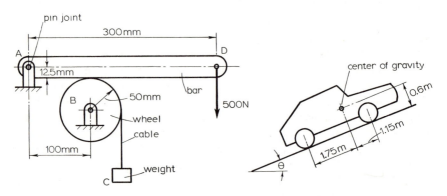

Fig. P4.1   **Fig. P4.2**

4.5 A 20-kg uniform ladder rests against a wall as shown in Fig. P4.5. Determine the maximum height of the rung above the ground to which a 70-kg person can climb without the ladder slipping. The coefficient of static friction between the ladder and the floor is 0.2 and between the ladder and the wall is 0.3.

4.6 The total mass of a grocery cart and its contents is 50 kg (Fig. P4.6). What force F_p would be required to initiate motion of the cart in the direction of the force if (a) the front wheels are locked but the rear wheels are free to rotate, (b) the rear wheels are locked but the front wheels are free to rotate, and (c) all wheels are locked? The coefficient of static friction between the wheels and the floor is 0.3.

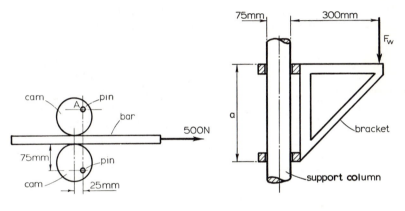

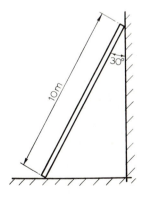

Fig. P4.3   **Fig. P4.4**

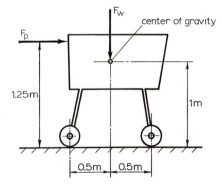

Fig. P4.5   **Fig. P4.6**

4.7   Two homogeneous crates of mass 25 kg and 15 kg are resting on top of each other as shown in Fig. P4.7. What is the largest force F_p that can be applied to the top crate so that no motion of either crate takes place? The coefficient of static friction between the two crates is 0.4, while the coefficient of static friction between the lower crate and the floor is 0.35.

4.8   What minimum horizontal force F_p must be applied to the drawer shown in Fig. P4.8 to push it in? The drawer and its contents weigh 100 N. The coefficient of static friction between all surfaces is 0.3.

4.9   A gravity feed track used for feeding long cylindrical parts consists of two sides at right angles to each other as shown in Fig. P4.9. If the coefficient of static friction between the parts and the track is 0.60, what is the minimum allowable angle θ of inclination of the track so that the parts will always slide down?

4.10   The automobile of weight F_w is parked on a hill with brakes applied to the rear wheels only (Fig. P4.10). Obtain expressions for the components of the

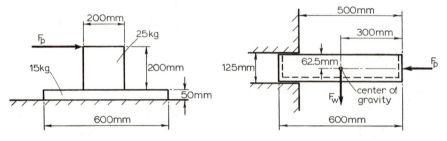

Fig. P4.7   **Fig. P4.8**

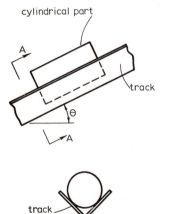

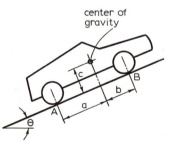

Fig. P4.9   **Fig. P4.10**

reactions on each wheel at *A* and *B* parallel and normal to the road, respectively.

4.11 Figure P4.11 shows a "dust bug" attachment which is mounted on pivots on the head of a record player tone arm. It is required to design this dust bug attachment so that when it is in use it does not affect the vertical (tracking) force on the stylus. If the coefficient of dynamic friction between the brush and the record surface is 0.2, what distance *x* (millimetres) should the center of gravity of the attachment overhang the brush?

4.12 In expensive record players, an adjustment is provided known as the "antiskating" adjustment. This takes the form of a mechanism or spring which counteracts the tendency of the stylus to "skate" radially on the record surface. For the arrangement shown in Fig. P4.12, what force F_p should the antiskating device exert at the commencement of playing a 12-in. record? Assume that the effective coefficient of friction between the stylus and the record surface is 0.2 and that the vertical "tracking" force on the stylus is 10 mN.

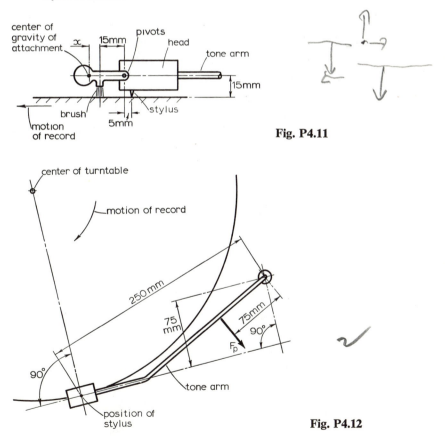

Fig. P4.11

Fig. P4.12

4.13 The concrete slab shown in Fig. P4.13 is 12 m long and weighs 2 kN/m. It is to be pushed up the step by a horizontal force F_p. If the static coefficient of friction at points A and B is 0.5, determine the magnitude of F_p so that the slab starts to move.

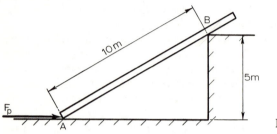

Fig. P4.13

5

Distributed Forces

5.1 Introduction

In reality most forces are distributed over an area rather than applied at a single point. The assumption of point loading is usually valid in those cases where the area over which the force is applied is small in comparison with the size of the body. If this is not the case, then the distribution of the forces must be accounted for. In practice, distributed forces or distributed loads will arise because of wind pressures on structures, hydrostatic (fluid) pressures on submerged objects, or snow accumulation on buildings or simply because of the distribution of mass throughout an object or structural member.

5.2 Distributed Loading

Figure 5.1 shows a graph of a distributed load on a narrow beam where the instantaneous value of the mass of the load per unit length of the beam is m_l. Thus, the force due to the load acting on the small element shown is $m_l g\, dx$, and the total force is obtained by summing the forces on all such elements for values of x lying between a and b. Mathematically, therefore, the total force F_w is given by

$$F_w = \int_a^b m_l g\, dx \qquad (5.1)$$

In statics problems it is often useful to know the point at which the total force F_w may be assumed to act on a rigid body so that the moment of this

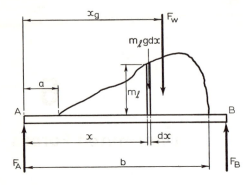

Fig. 5.1 Distributed load on narrow beam.

force about any point is the same as the moment due to the distributed load. In the present example, this point will be assumed to be a distance x_g from the left-hand end of the beam A. Thus, the moment of the equivalent force F_w about A must be equal to the sum of the moments of the weights of the individual elements of the distributed load. A convenient way to remember this procedure is to state that "the moment of the sum equals the sum of the moments." Mathematically,

$$F_w x_g = \int_a^b m_l g x \, dx$$

or

$$x_g = \frac{\int_a^b m_l g x \, dx}{\int_a^b m_l g \, dx} \tag{5.2}$$

Example 5.1   For the linearly increasing loading on a narrow beam shown in Fig. 5.2, determine the total weight of the load F_w and the distance x_g from A where this total weight can be assumed to act.

Solution.   Before using Eqs. (5.1) and (5.2) it is necessary to obtain the relationship between m_l and x. For the linearly increasing load shown in Fig. 5.2 we can write

$$m_l = \frac{x}{l} m_B$$

Also, since $a = 0$ and $b = l$, Eq. (5.1) becomes

$$F_w = \int_o^l \frac{x}{l} m_B g \, dx$$

$$= \frac{m_B g}{l} \left(\frac{x^2}{2} \right)_o^l = \frac{m_B g l}{2}$$

It should be noted that this result for F_w is equal to the area of the graph of the loading.

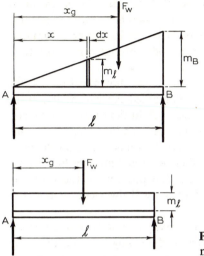

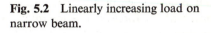

Fig. 5.2   Linearly increasing load on narrow beam.

Fig. 5.3   Uniformly distributed load on narrow beam.

Equation (5.2) becomes

$$x_g = \frac{\int_o^l (m_B g/l)x^2 \, dx}{m_B g(l/2)} = \frac{2}{3}l$$

It will be seen later that this value for x_g is equal to the distance to the centroid of the area of the graph of the loading.

Example 5.2   For the uniformly distributed load on a narrow beam shown in Fig. 5.3, determine its total weight and the point at which this total weight may be assumed to act.

Solution.   With a uniformly distributed load, it can be seen by inspection that the total weight can be assumed to act in the center of the distributed load. Alternatively, from Eq. (5.1),

$$F_w = m_l g l$$

while from Eq. (5.2),

$$x_g = \frac{l}{2}$$

5.3   Center of Gravity

Another example of a distributed force is the action of gravity on a body. Since gravity applies a force to each individual particle in a body, the weight of the body should be represented by many small forces distributed through-out the entire body. These forces, however, can always be reduced to a single force acting through a point within the body called its center of gravity. Thus, the center of gravity is the point in a body through which the total weight of the body can be assumed to act.

For bodies of uniform thickness such as the shape shown in Fig. 5.4, the position of the center of gravity can easily be obtained experimentally. If the body is suspended by a thread, then the tension F_t in the thread will be col-linear with the force mg due to gravity, and its line of action will pass through the center of gravity of the body. If this line of action is drawn on the body and then the body is suspended from another point and another line drawn, the intersection of the two lines will define the position of the center of gravity. Of course, the center of gravity of a body will always lie on any of its axes of symmetry.

To determine the position of the center of gravity of a body by analysis, the same basic procedure is used as for distributed loading on a beam. If the center of gravity of a body lies at a distance x_g from an arbitrarily chosen axis, then the moment of the total weight $F_w x_g$ of the body about that axis must be equal to the sum of the moments of the weights of the elements of the body about that same axis; that is,

$$F_w x_g = \int m_l g x \, dx \tag{5.3}$$

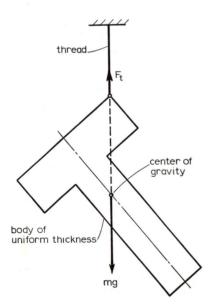

Fig. 5.4 Experimental method for ob-taining the position of the center of grav-ity of a body of uniform thickness.

It should be noted that we are again applying the statement that the moment of the sum equals the sum of the moments.

For many bodies, the position of the center of gravity can be determined by dividing the body into portions for each of which the position of the center of gravity is known. In this case Eq. (5.3) can be rewritten in the form

$$F_w x_g = m_1 g x_{g1} + m_2 g x_{g2} + \cdots \tag{5.4}$$

where x_{g1}, x_{g2}, etc., are the distances to the axis under consideration for the individual portions having masses m_1, m_2, etc., respectively.

Alternatively, if γ is the specific weight (weight per unit volume) of a body of uniform density and of uniform thickness h,

$$F_w = \gamma h \int dA \tag{5.5}$$

and
$$F_w x_g = \gamma h \int x \, dA \tag{5.6}$$

where dA is a small element of area of the body and x is the distance of this element to the axis under consideration.

Now, substitution of Eq. (5.5) into Eq. (5.6) gives

$$x_g \int dA = x_g A = \int x \, dA \tag{5.7}$$

and thus for a body of uniform density and thickness that can be divided conveniently into individual portions for which the location of the center of gravity of each is known, we can write

$$x_g A = x_{g1} A_1 + x_{g2} A_2 + \cdots \tag{5.8}$$

Example 5.3   Determine the location of the center of gravity for the plane body of uniform density shown in Fig. 5.5.

Solution.   The body can be divided into two rectangles (1) and (2). By symmetry, the center of gravity for each rectangle will lie in the center of the rectangle, and, clearly, the center of gravity of the whole body must lie on the axis of symmetry x. Now, from Eq. (5.8),

$$\text{moment of sum} = \text{sum of moments}$$

$$x_g(ac + bd) = \frac{a^2 c}{2} + \frac{b^2 d}{2} + abd$$

or
$$x_g = \frac{a^2 c + b^2 d + 2abd}{2(ac + bd)}$$

If a plane body is asymmetrical, then a second calculation must be performed to determine the distance of the center of gravity from the x axis.

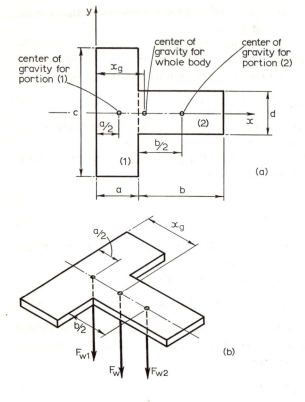

Fig. 5.5 Symmetrical plane body of uniform density.

Example 5.4 Determine the location of the center of gravity for the plane body of uniform density shown in Fig. 5.6a.

Solution. In this case, the body can be divided into the two rectangles shown in Fig. 5.6b. Now, from Eq. (5.8),

$$\text{moment of sum} = \text{sum of moments}$$

About the y axis: $x_g(ac + bd) = \dfrac{a^2c}{2} + \dfrac{b^2d}{2} + abd$

About the x axis: $y_g(ac + bd) = \dfrac{ac^2}{2} + \dfrac{bd^2}{2}$

or

$$x_g = \frac{a^2c + b^2d + 2abd}{2(ac + bd)}$$

$$y_g = \frac{ac^2 + bd^2}{2(ac + bd)}$$

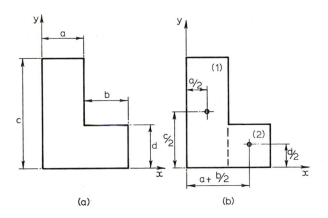

Fig. 5.6   Asymmetrical plane body.

5.4   Centroid of Area

For plane bodies of uniform density such as those considered in the previous examples, the center of gravity lies at the centroid of the plane area. The centroid refers only to the geometrical shape of the body, and thus for bodies of nonuniform density the center of gravity is displaced from the centroid of area.

If we now define the coordinates of the centroid of area as (x_a, y_a), Eqs. (5.7) and (5.8) become

$$x_a A = \int x \, dA \tag{5.9}$$

$$x_a A = x_{a1} A_1 + x_{a2} A_2 + \cdots \tag{5.10}$$

and similarly,

$$y_a A = \int y \, dA \tag{5.11}$$

$$y_a A = y_{a1} A_1 + y_{a2} A_2 + \cdots \tag{5.12}$$

5.5   Center of Mass

The center of mass of a body is defined as the point in the body where the entire mass of the body can be considered to act so that

$$m x_c = m_1 x_1 + m_2 x_2 + \cdots \tag{5.13}$$

where m is the total mass of the body and $m_1, m_2, \ldots$ are the masses of the elements which make up the body. If Eq. (5.13) is multiplied by g, then it becomes identical to Eq. (5.4), and any distinction between center of mass and center of gravity appears to be unnecessary. For most engineering situations where the bodies are within a uniform, parallel gravitational field, the centers of mass and of gravity do coincide. Only in the case of large slender bodies orbiting the earth, for example, where the body is contained in a nonparallel gravitational field, is there a significant difference between the positions of the centers of mass and of gravity.

5.6   Center of Gravity of a Three-Dimensional Body

The bodies considered so far have been bodies of uniform thickness whose center of gravity has been located halfway between the two plane faces. Thus, only two independent calculations were needed to determine the coordinates (x_g, y_g) of the center of gravity in the xy plane. For a general three-dimensional body, an additional calculation is required in order to locate the center of gravity along the z axis perpendicular to the xy plane.

If the body has a uniform density, then the center of gravity and centroid of volume coincide, and equations similar to those for a plane area [Eqs. (5.9)–(5.12)] can be written as follows:

$$V x_v = \int x \, dV \tag{5.14}$$

$$V x_v = V_1 x_{v1} + V_2 x_{v2} + \cdots \tag{5.15}$$

Similar equations can be written for y_v and z_v.

Thus, the coordinates of the centroid of volume are (x_v, y_v, z_v).

Example 5.5   Determine the location of the centroid of volume for the right circular solid cone of height h shown in Fig. 5.7.

Solution.   Because of symmetry, y_v and z_v are equal to zero, and it is only necessary to determine the coordinate x_v. Upon dividing the body into disk

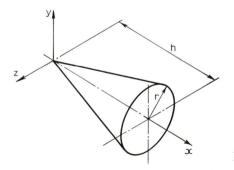

Fig. 5.7   Solid cone.

elements (Fig. 5.8) of radius y and thickness dx, from Eq. (5.14),

$$x_v = \frac{\int x\, dV}{V} = \frac{\int_o^h x\pi y^2\, dx}{\int_o^h \pi y^2\, dx} \tag{a}$$

Since

$$y = \frac{rx}{h} \tag{b}$$

then

$$x_v = \frac{\int_o^h x^3\, dx}{\int_o^h x^2\, dx}$$
$$= \frac{3h}{4}$$

Example 5.6   Determine the location of the centroid of volume for the right circular cone with a concentric circular cylindrical hollow in its base as shown in Fig. 5.9.

Solution.   Again, because of symmetry, y_v and z_v are equal to zero, and it is only necessary to determine x_v. The body can be considered a solid cone with

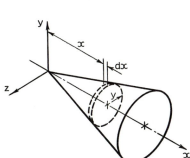

Fig. 5.8 Construction for the solution of Example 5.5.

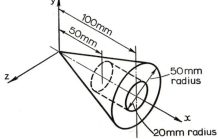

Fig. 5.9 Cone with cylindrical hollow.

a cylinder removed, as shown in Fig. 5.10. Thus,

$$Vx_v = V_1 x_{v1} - V_2 x_{v2} \qquad \text{(a)}$$

where
$$V_1 = \frac{1}{3}\frac{\pi}{4}(100)^2(100) = 261.8 \times 10^3 \text{ (mm)}^3$$

$$V_2 = \frac{\pi}{4}(40)^2(50) = 62.8 \times 10^3 \text{ (mm)}^3$$

$$V = V_1 - V_2 = 199 \times 10^3 \text{ (mm)}^3$$

Substitution of these values into Eq. (a) gives

$$x_v = 75 \text{ mm}$$

This result could have been deduced from Fig. 5.10. Since the centroid of volume of the cylindrical hole is coincident with the centroid of volume of the solid cone, subtracting the former from the latter has no effect on the position of the centroid of volume.

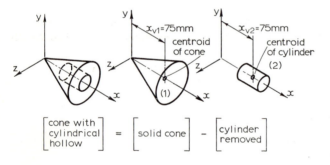

Fig. 5.10 Construction for solution of Example 5.6.

Problems

5.1   For the parabolically distributed load on a narrow beam shown in Fig. P5.1, determine the total weight F_w of the load and the distance x_g from A to the point where this total load can be assumed to act.

5.2   For the distributed load on a beam shown in Fig. P5.2, determine the total load F_w and the distance x_g from A to the point where this total load can be assumed to act.

5.3   Determine the location of the center of gravity for the plane body of uniform density and uniform thickness shown in Fig. P5.3.

5.4   Determine the location of the centroid of area for the plane area shown in Fig. P5.4.

5.5 Determine the coordinates (x_g, y_g) of the center of gravity of the segment of thin wire shown in Fig. P5.5.

5.6 Determine the location of the center of gravity of the thin wire shape shown in Fig. P5.6.

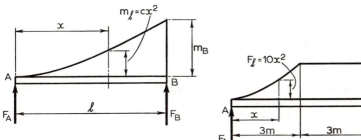

$m_\ell = cx^2$

c = constant

Fig. P5.1

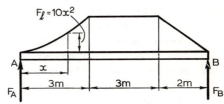

$F_\ell = 10x^2$

3m   3m   2m

Fig. P5.2

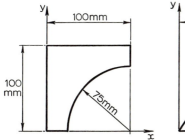

100mm

100 mm

75mm

Fig. P5.3

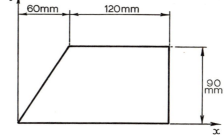

60mm   120mm

90 mm

Fig. P5.4

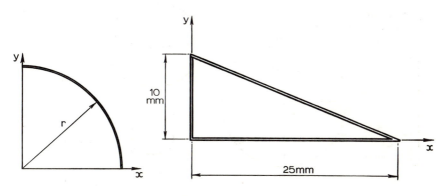

r

10 mm

25mm

Fig. P5.5   **Fig. P5.6**

5.7   Determine the location of the center of gravity of the disk and pin assembly
      shown in Fig. P5.7. Note: The density of steel is 7200 kg/m³ and of aluminum
      2400 kg/m³.

5.8   Determine the location of the center of gravity of the headed component
      shown in Fig. P5.8.

5.9   Determine the depth l of the concentric hole in the component shown in
      Fig. P5.9 such that the center of gravity of the component lies a minimum
      distance from the end opposite that of the hole.

5.10 The block of cement shown in Fig. P5.10 has a uniform thickness and a mass
      of 1500 kg. At what minimum height h must the force F_p be applied to the
      bracket in order to turn the block over point A without the block slipping?
      What is the magnitude of F_p under these conditions? Assume that the coeffi-
      cient of static friction between the block and the floor is 0.4.

5.11 A uniform shaft with two protruding arms is shown in Fig. P5.11. Locate the
      center of gravity of the system if each uniform arm weighs 25 N and the
      shaft weighs 175 N.

5.12 Determine the magnitudes of the reactions F_A and F_B for the arrangements
      shown in Fig. P5.12.

5.13 Determine the coordinates of the center of mass for the plane bodies shown
      in Fig. P5.13.

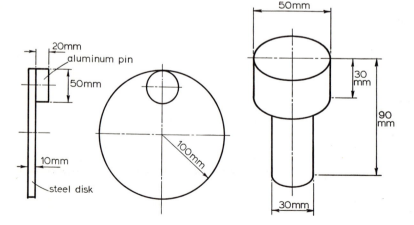

Fig. P5.7   **Fig. P5.8**

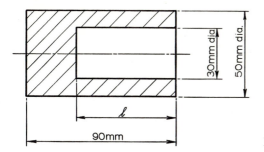

Fig. P5.9

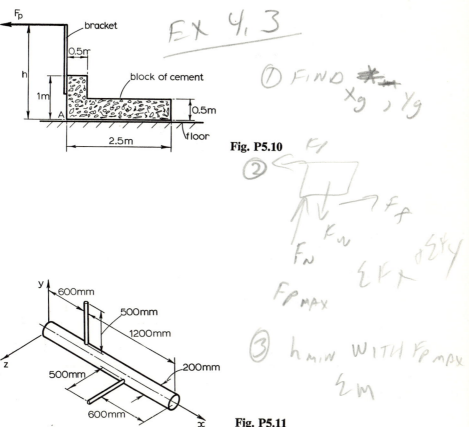

bracket

0.5m

block of cement

h

1m

A

2.5m

floor

0.5m

Fig. P5.10

Fig. P5.11

y

600mm

500mm

1200mm

z

200mm

500mm

600mm

x

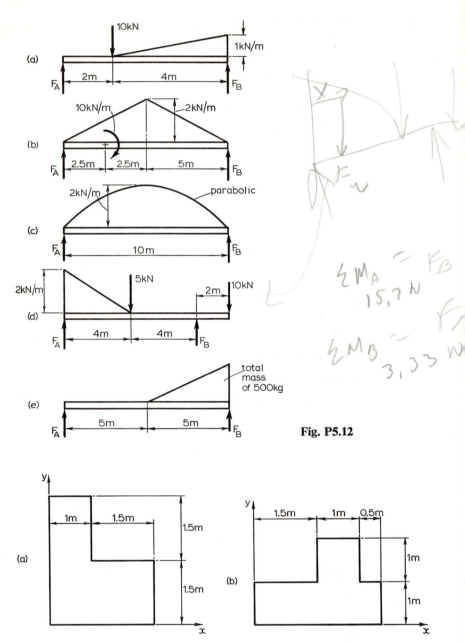

Fig. P5.12

Fig. P5.13

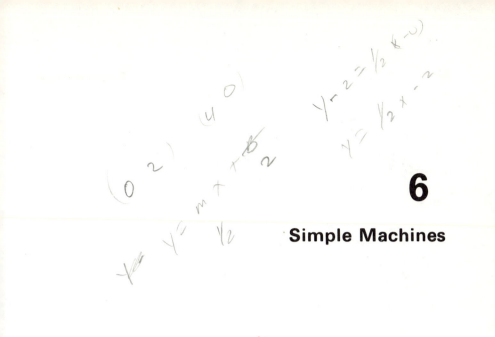

6

Simple Machines

6.1 Introduction

The previous chapters have been devoted primarily to the presentation of the general theory of statics of particles and rigid bodies subjected to coplanar forces. The next several chapters are devoted to an examination of some important engineering applications of the theory already considered.

6.2 Simple Machines

A "simple" machine is a device used to increase force, to increase speed, or to change the direction of a force. There are only six simple machines: lever, pulley, wheel and axle, inclined plane, wedge, and screw.

6.3 Lever

A lever is a rigid bar, straight or curved, with a pivot or fulcrum at some point along its length so that a force acting at one point can apply a much larger force at a different point along the lever. Levers are frequently used to move heavy loads or to operate mechanisms.

Figure 6.1 illustrates three types of levers. Point O is the pivot or fulcrum, F_l is the load force, F_p is the applied force, and the distances a and b are the

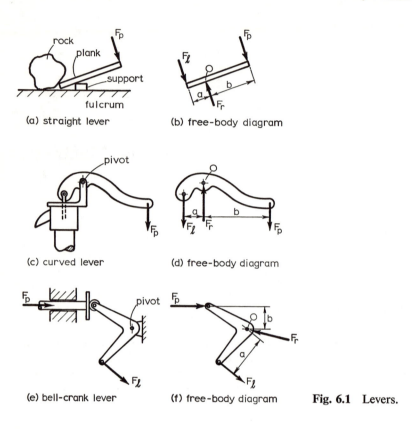

(a) straight lever

(b) free-body diagram

(c) curved lever

(d) free-body diagram

(e) bell-crank lever

(f) free-body diagram

Fig. 6.1 Levers.

moment arms from the lines of action of the forces to the pivot or fulcrum and are sometimes called lever arms. For static equilibrium about the pivot or fulcrum, the sum of the moments must be zero so that

$$\sum M_O = 0: \qquad\qquad aF_l = bF_p$$

or

$$\frac{F_l}{F_p} = \frac{b}{a} \qquad\qquad (6.1)$$

The ratio b/a is called the mechanical advantage of the lever. If $b/a > 1$, then the magnitude of the force F_p required to balance the load force F_l will be less than F_l, but the displacement of the point of application of F_p will be greater than the displacement of the point of application of F_l.

Example 6.1 If the mechanical advantage of the curved lever shown in Fig. 6.1c is 3.5, what force F_p is required to raise 8 kg of water?

Solution. In this case the magnitude of the load F_l is given by

$$F_l = mg = 8(9.81) = 78.5 \text{ N}$$

Thus, from Eq. (6.1),

$$F_p = \frac{a}{b}F_l = \frac{1}{3.5}(78.5) = 22.4 \text{ N}$$

6.4 Pulley

A pulley consists of a wheel which is free to turn on its axle and carries a rope, cable, or belt. For a system consisting of a single pulley, one end of the rope is attached to the load while a pull of F_p is applied to the other end. Figure 6.2a shows a single pulley which is being used to change the direction of the applied force in order to raise a load of mass m. The free-body diagrams for the pulley and the mass are shown in Fig. 6.2b. Considering static equilibrium of the mass gives

$$\sum F_y = 0: \qquad\qquad F_{r3} - F_w = 0 \qquad\qquad (6.2)$$

Considering static equilibrium of the pulley gives

$$\sum F_x = 0: \qquad\qquad F_{r1} + F_p \cos \theta = 0 \qquad\qquad (6.3)$$

$$\sum F_y = 0: \qquad\qquad F_{r2} - F_p \sin \theta - F_{r3} = 0 \qquad\qquad (6.4)$$

$$\sum M_O = 0: \qquad\qquad F_{r3}r - F_p r = 0 \qquad\qquad (6.5)$$

Solving these equations gives

$$F_{r3} = F_w \qquad\qquad (6.6)$$

$$F_p = F_w \qquad\qquad (6.7)$$

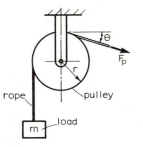

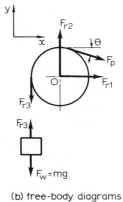

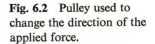

(a) geometry
(b) free-body diagrams

Fig. 6.2 Pulley used to change the direction of the applied force.

$$F_{r1} = -F_p \cos \theta = -F_w \cos \theta \tag{6.8}$$

$$F_{r2} = F_w(1 + \sin \theta) \tag{6.9}$$

Thus, when the pulley is in static equilibrium and no friction arises at the axle, the tension in the (massless) cable is constant and is transmitted around the pulley to raise the load.

Figure 6.3a shows a system consisting of two pulleys. From the free-body diagrams shown in Fig. 6.3b, for pulley A we have

$$\sum M_{OA} = 0: \qquad\qquad F_{r2} = F_p \tag{6.10}$$

while for pulley B, if its weight is neglected,

$$\sum M_{OB} = 0: \qquad\qquad F_{r3} = F_{r2} \tag{6.11}$$

$$\sum F_y = 0: \qquad\qquad F_{r3} + F_{r2} = F_w \tag{6.12}$$

Combining Eqs. (6.10)–(6.12) gives

$$\frac{F_w}{F_p} = 2 \tag{6.13}$$

The last equation shows that a load F_w equal to twice the magnitude of the applied force F_p can be raised with this system; that is, the mechanical advantage of this pulley system is two.

Example 6.2 A system of four pulleys is shown in Fig. 6.4a. Determine the force F_p required to raise the load F_w. Neglect friction and the weights of the pulleys and ropes.

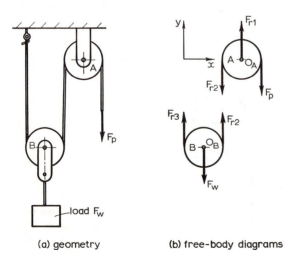

(a) geometry   (b) free-body diagrams

Fig. 6.3 Two-pulley system.

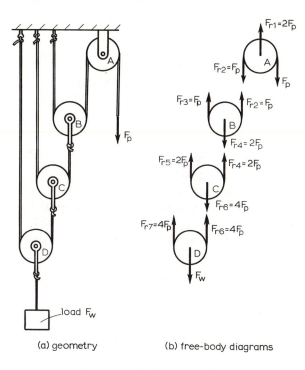

(a) geometry   (b) free-body diagrams

Fig. 6.4   Pulley system for Example 6.2.

Solution.   Figure 6.4b shows the free-body diagrams for the four pulleys. For pulley A, summing moments about its axis gives

$$F_{r2} = F_p \tag{a}$$

Similarly, for pulley B,

$$F_{r3} = F_{r2} = F_p \tag{b}$$

Summing forces in the vertical direction for pulley B gives

$$F_{r3} + F_{r2} - F_{r4} = 0 \tag{c}$$

or $$F_{r4} = F_{r3} + F_{r2} = 2F_p \tag{d}$$

Following a similar procedure for pulley C gives

$$F_{r6} = 4F_p \tag{e}$$

Hence, summing forces in the vertical direction for pulley D gives

$$2F_{r6} - F_w = 0$$

$$2(4)F_p = F_w$$

$$F_p = \frac{F_w}{8} \tag{f}$$

Therefore, the mechanical advantage for this pulley system is eight.

6.5 Wheel and Axle

A wheel and axle consists of a wheel attached to a smaller-diameter shaft or axle which turns with the wheel. A door knob, a steering wheel, and a screwdriver bit in a brace are common examples of a wheel and axle. Figure 6.5 shows a free-body diagram for a wheel and axle. The force applied to the wheel is denoted by F_p, while the load force applied to the shaft is denoted by F_l. For static equilibrium, the sum of the moments about the axis of the wheel and axle must be zero, so that

$$F_l a = F_p b$$

or $$\frac{F_l}{F_p} = \frac{b}{a} \tag{6.14}$$

Therefore, the ratio b/a is the mechanical advantage of the wheel and axle.

Example 6.3 For the arrangement shown in Fig. 6.6a, determine the minimum force required on the handle of the crank to raise a 10-kg bucket of water at a constant speed.

Solution.   From the free-body diagram shown in Fig. 6.6b,

$$\sum M_O = 0: \qquad\qquad F_p(200) = 98.1(100)$$

$$F_p = 49.05 \text{ N}$$

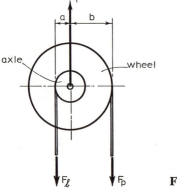

Fig. 6.5 Wheel and axle.

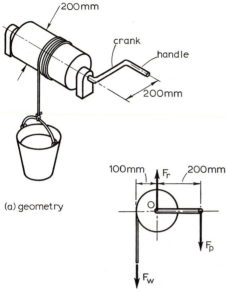

200mm

crank

handle

200mm

(a) geometry

100mm $\uparrow F_r$ 200mm

$\downarrow F_p$

$\downarrow F_w$

(b) free-body diagram

Fig. 6.6 Wheel and axle arrangement for Example 6.3.

6.6 Inclined Plane

An inclined plane (Fig. 6.7a) can be used to raise a large mass through a small vertical distance by pushing the mass up the plane.

To determine the force F_p required to move the mass having a weight F_w up the inclined plane, the free-body diagram for the mass (Fig. 6.7b) is utilized. Resolving forces in the x and y directions gives

$$\sum F_x = 0: \qquad F_p - F_w \sin \theta - \mu F_n = 0 \qquad (6.15)$$

$$\sum F_y = 0: \qquad F_n - F_w \cos \theta = 0 \qquad (6.16)$$

where μ is the appropriate value of the coefficient of friction. Solving these equations for F_p gives

$$F_p = F_w(\sin \theta + \mu \cos \theta) \qquad (6.17)$$

$$F_p = \frac{F_w}{l}(h + \mu b) \qquad (6.18)$$

The mechanical advantage in this case is given by

$$\frac{F_w}{F_p} = \frac{l}{h + \mu b} \qquad (6.19)$$

The value of μ can sometimes be reduced by placing the mass on rollers.

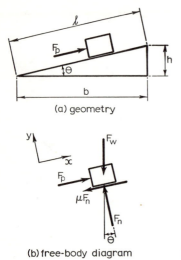

(a) geometry

(b) free-body diagram   **Fig. 6.7**   Inclined plane.

6.7 Wedge

A wedge is shown in Fig. 6.8 and is a special application of the inclined plane; it can be regarded as two inclined planes set base to base. Knife blades, axe blades, and chisels are examples of wedges and can be used to cut or split material. Driving a wedge into the material to be cut or split can produce large forces tending to separate the two sections. If friction is neglected, then for equilibrium

$$F_p = 2F_n \sin \theta = \frac{2h}{l} F_n \qquad (6.20)$$

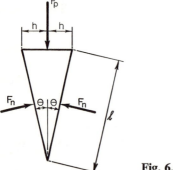

Fig. 6.8   Wedge.

The mechanical advantage is given by

$$\frac{F_n}{F_p} = \frac{l}{2h} \qquad (6.21)$$

6.8 Screw

A screw (Fig. 6.9) is sometimes described as an inclined plane wrapped around a cylinder.* By using a screw, it is possible to generate a large force along the axis of the screw by applying a relatively small moment to the screw. Since friction plays an important role in the use of screws, the derivation of the force and moment relationships will be presented later in Chapter 10, which deals with problems involving friction.

*To show this, cut a sheet of paper in the shape of a right triangle and wind it around a pencil so that its hypotenuse forms a helical thread.

Fig. 6.9   Screw.

Problems

6.1   Figure P6.1 shows an airplane rudder pedal connected to a cable. If a force of 1250 N is required in the cable during a certain maneuver, what force F_p must be applied to the pedal? What is the mechanical advantage of the pedal?

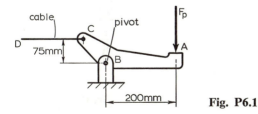

Fig. P6.1

6.2   A force of 150 N is applied to handle *AC* of the pump shown in Fig. P6.2. Determine the force transmitted to the piston for the position shown. What is the mechanical advantage of the system?

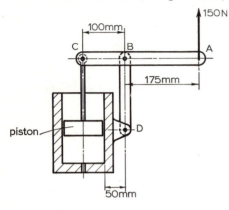

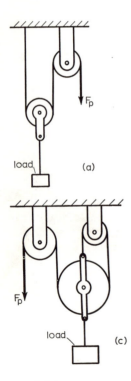

Fig. P6.2

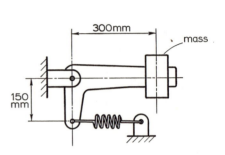

Fig. P6.3

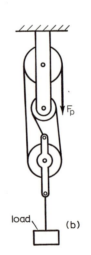

Fig. P6.4

Fig. P6.5

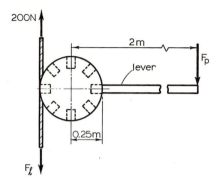

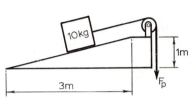

Fig. P6.6

Fig. P6.7

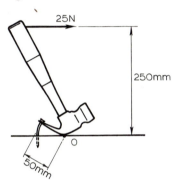

Fig. P6.8

6.3 Figure P6.3 shows a type of seismic instrument used to record vibrations. What spring force is required to hold the 1-kg mass in the position shown? Neglect the weight of the lever.

6.4 For the pulley arrangements shown in Fig. P6.4, what force F_p is required to raise a 200-kg load? Neglect friction. What is the mechanical advantage of each system?

6.5 The screwdriver bit in a brace shown in Fig. P6.5 is used to drive a tight screw. If a force F_p of 50 N is applied to the handle in a direction perpendicular to the plane of the drawing, determine the magnitude of the moment or torque applied to the screw.

6.6 A rope is coiled around an 0.5-m-diameter ship's capstan (Fig. P6.6). One end of the rope is secured to the dock; the other end is held by a sailor, who is applying a force of 200 N. What force F_p must be applied to the capstan lever to withstand a tension of 10 kN in the section of the rope between the ship and the dock?

6.7 Determine the force F_p required to initiate motion of the load up the plank shown in Fig. P6.7. The coefficient of static friction between the load and the plank is 0.4. Neglect friction in the pulley.

6.8 A 25-N force is applied to the hammer shown in Fig. P6.8. If the hammer pivots about point O, determine the force exerted on the head of the nail.

7

Trusses and Frames

7.1 Introduction

Frames (Fig. 7.1a) and trusses (Fig. 7.1b) are engineering structures designed to support loads. Frames and trusses are made up of structural members such as bars, I-beams, and channels which are fastened together by welding or riveting or with large bolts or pins. A frame can be designed to support both moments and forces applied at any point in the structure. A truss, however, can be regarded as a pin-jointed frame made up of relatively slender members and is designed primarily to support forces applied at the joints. Trusses are used in structures such as roofs, bridges, antenna towers, and roller-coaster rides at amusement parks.

While many trusses are formed by riveting or welding the ends of the mem-

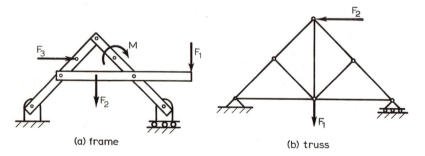

(a) frame   (b) truss

Fig. 7.1 Trusses and frames.

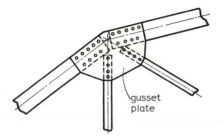

Fig. 7.2 A riveted truss joint where the centerlines of the members are concurrent.

bers to gusset plates (Fig. 7.2), the assumption of a pin-jointed connection is usually satisfactory if the centerlines of the members are concurrent at the joint.

7.2 Analysis of Trusses

In the determination of the forces in the various members of a truss, it is assumed that each member is pin-jointed at its ends. This means that the lines of action of the two forces applied to the ends of a member in a truss must lie along the straight line joining the ends of the member. When these two forces tend to compress the member, the member is said to be in compression and is known as a "strut" (Fig. 7.3a). When these two forces tend to stretch the member, the member is said to be in tension and is known as a "tie" (Fig. 7.3b). Long, slender members which are designed to accept a certain tensile force will often be incapable of taking a compressive force of the same magnitude without buckling. For example, consider a mass suspended from a vertical thin wire. Now if the mass were to be placed on top of the wire, the wire would bend and buckle. It is therefore important in analyzing trusses to determine whether a particular member is in tension or compression.

The methods described here apply only to trusses that can be built up by starting with a triangular truss consisting of three members and then further triangular trusses formed by adding two members at a time.

F_p → ○————————○ ← F_p

(a) member in compression (strut)

← F_p ○————————○ F_p →

(b) member in tension (tie)

Fig. 7.3 Members of a truss in tension and compression.

7.2.1   Graphical Method (Maxwell Diagram)

The procedure is to draw a vector diagram for the external forces acting on the structure and then add to this the vector diagrams for the forces acting at each junction of the members (joint) in the truss as follows:

1. Letter the spaces in the framework and the spaces between each external force (Fig. 7.4a). This is known as Bow's notation.
2. Determine the reactions at the supports (Fig. 7.4b).
3. Draw a vector diagram for the external forces, moving counterclockwise around the frame; see line *abca* in Fig. 7.4c.
4. Add the vector diagrams for the forces at each joint. In the present ex-

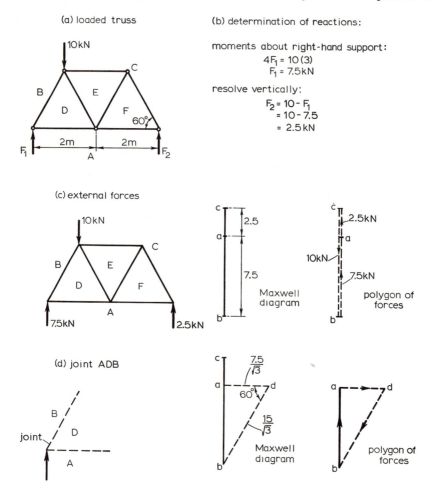

(a) loaded truss

(b) determination of reactions:

moments about right-hand support:
$$4F_1 = 10\,(3)$$
$$F_1 = 7.5\,\text{kN}$$

resolve vertically:
$$F_2 = 10 - F_1$$
$$= 10 - 7.5$$
$$= 2.5\,\text{kN}$$

(c) external forces

Maxwell diagram

polygon of forces

(d) joint ADB

Maxwell diagram

polygon of forces

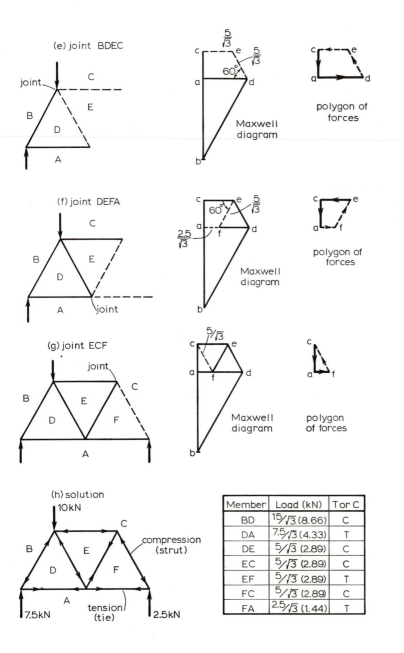

Fig. 7.4 Diagrams for the graphical method.

(e) joint BDEC

$\frac{5}{\sqrt{3}}$

$\frac{5}{\sqrt{3}}$

60°

Maxwell diagram

polygon of forces

(f) joint DEFA

60°

$\frac{5}{\sqrt{3}}$

$\frac{2.5}{\sqrt{3}}$

Maxwell diagram

polygon of forces

(g) joint ECF

$\frac{5}{\sqrt{3}}$

Maxwell diagram

polygon of forces

(h) solution

10 kN

compression (strut)

tension (tie)

7.5 kN

2.5 kN

Member	Load (kN)	T or C
BD	$\frac{15}{\sqrt{3}}$ (8.66)	C
DA	$\frac{7.5}{\sqrt{3}}$ (4.33)	T
DE	$\frac{5}{\sqrt{3}}$ (2.89)	C
EC	$\frac{5}{\sqrt{3}}$ (2.89)	C
EF	$\frac{5}{\sqrt{3}}$ (2.89)	T
FC	$\frac{5}{\sqrt{3}}$ (2.89)	C
FA	$\frac{2.5}{\sqrt{3}}$ (1.44)	T

ample, starting at the left-hand support, joint ADB, where two unknown forces occur in members BD and DA, draw a line through a parallel to DA and a line through b parallel to DB (Fig. 7.4d). Label the intersection of these two lines d. This completes the vector diagram for the joint ADB, and the forces in BD and DA can be determined by calculation or measurement of the lengths of the lines bd and ad. Considering joint $BDEC$ (Fig. 7.4e) where the forces in members CE and ED are unknown, draw a line through c parallel to CE and a line through d parallel to DE; label this intersection e. This completes the vector diagram for joint $BDEC$. Considering next the joint $DEFA$ (Fig. 7.4f), draw a line through e parallel to EF and a line through a parallel to AF. Finally, considering the joint ECF, draw a line through f parallel to CF (Fig. 7.4g), and the diagram is complete.

5. Determine the directions of the forces in the members at each joint in the truss and hence determine whether each member is in compression or tension. This is done by considering the polygon of forces for each joint and determining whether the end of the member pushes on the joint (compression) or pulls on the joint (tension). For example, from the polygon of forces shown in Fig. 7.4d it can be seen that the force exerted by member BD on the lower joint is directed down and to the left. Thus, member BD is pushing on the joint and is in compression. Similarly, the force in member DA is directed to the right, pulling on the same joint, and member DA is therefore in tension. The complete solution to the analysis is summarized in Fig. 7.4h. In practice, it is not necessary to draw all the individual diagrams shown in Fig. 7.4. In this example, the steps necessary to construct the final Maxwell diagram were all shown individually.

Example 7.1   Determine the force in each member of the truss shown in Fig. 7.5a.

Solution.   Resolving horizontally for the truss, we get

$$\sum F_x = 0: \qquad\qquad\qquad F_{1h} = 4 \text{ kN}$$

Taking moments about support 1 gives

$$\sum M_1 = 0: \qquad\quad 3F_2 = 1.5(6) + \frac{\sqrt{3}}{2}(1.5)(4) = 14.2$$

or

$$F_2 = \frac{14.2}{3} = 4.73 \text{ kN}$$

Resolving vertically, we get

$$\sum F_y = 0: \qquad\qquad\quad F_{1v} = 6 - 4.73 = 1.27 \text{ kN}$$

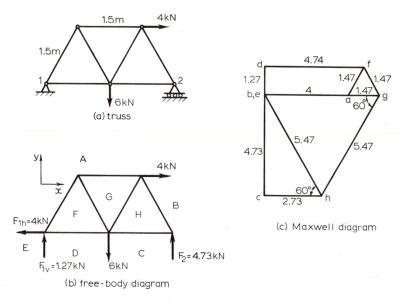

Fig. 7.5 Diagrams for Example 7.1.

The vector diagram for the external forces is now drawn starting at support 1 (*a* in the Maxwell diagram in Fig. 7.5c) and moving counterclockwise around the truss (lines *aedcba*). Starting at support 1 (joint *AEDF*), the vector diagrams for the forces at each joint in the truss are now added. The completed Maxwell diagram is shown in Fig. 7.5c, and the solution to the problem is as follows:

Member	Force (kN)	C or T
HB	5.47	C
GH	5.47	T
FG	1.47	T
HC	2.73	T
AG	1.47	C
AF	1.47	C
FD	4.74	T

7.2.2 Method of Joints

The analysis of the truss shown in Fig. 7.4a could also have been carried out by writing the equilibrium equations for each joint in the framework and solving for the unknown forces. For example, considering the free-body

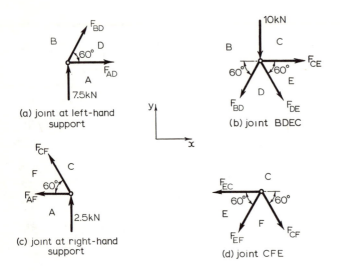

Fig. 7.6 Method of joints.

diagram for the joint at the left-hand support (Fig. 7.6a) and resolving horizontally, we get

$$\sum F_x = 0: \qquad\qquad F_{BD} \cos 60° + F_{AD} = 0 \qquad\qquad (7.1)$$

Resolving vertically gives

$$\sum F_y = 0: \qquad\qquad 7.5 + F_{BD} \sin 60° = 0 \qquad\qquad (7.2)$$

From Eq. (7.2),

$$F_{BD} = -\frac{7.5}{\sin 60°} = -8.66 \text{ kN}$$

From Eq. (7.1),

$$F_{AD} = -F_{BD} \cos 60° = \frac{-(-8.66)}{2} = 4.33 \text{ kN}$$

The negative sign for F_{BD} indicates that the force was shown in the wrong direction in Fig. 7.6a and that the member is in compression.

With the force in member *BD* known, we can now proceed to joint *BDEC* where the forces in members *CE* and *DE* are unknown. Considering the free-body diagram for this joint (Fig. 7.6b) and resolving horizontally, we get

$$\sum F_x = 0: \qquad -F_{BD} \cos 60° + F_{CE} + F_{DE} \cos 60° = 0 \qquad (7.3)$$

or, upon substitution for F_{BD},

$$F_{CE} + F_{DE} \cos 60° = (-8.66) \cos 60° = -4.33 \text{ kN} \qquad (7.4)$$

Resolving vertically, we get

$$\sum F_y = 0: \qquad -10 - F_{BD} \sin 60° - F_{DE} \sin 60° = 0 \qquad (7.5)$$

or substituting for F_{BD} and solving for F_{DE}, we get

$$F_{DE} = \frac{-10 - (-8.66) \sin 60°}{\sin 60°} = -2.89 \text{ kN}$$

Hence, from Eq. (7.4),

$$F_{CE} = -4.33 - (-2.89) \cos 60° = -2.89 \text{ kN}$$

Once again the negative signs for F_{DE} and F_{CE} indicate that the forces were shown in the wrong direction and that therefore both members are in compression.

The remaining unknown forces are those in members CF, EF, and AF, and these can be determined by considering free-body diagrams for the right-hand support (Fig. 7.6c) and joint CFE (Fig. 7.6d). Considering the free-body diagram for the right-hand support first (Fig. 7.6c) and resolving horizontally, we get

$$\sum F_x = 0: \qquad -F_{CF} \cos 60° - F_{AF} = 0 \qquad (7.6)$$

Resolving vertically gives

$$\sum F_y = 0: \qquad F_{CF} \sin 60° + 2.5 = 0 \qquad (7.7)$$

From Eq. (7.7),

$$F_{CF} = -2.89 \text{ kN}$$

while from Eq. (7.6),

$$F_{AF} = -F_{CF} \cos 60° = -(-2.89) \cos 60° = 1.44 \text{ kN}$$

Proceeding now to joint CFE and considering the free-body diagram shown in Fig. 7.6d, we can resolve forces in the vertical direction to finally get

$$\sum F_y = 0: \qquad -F_{EF} \sin 60° - F_{CF} \sin 60° = 0 \qquad (7.8)$$

or

$$F_{EF} = -F_{CF} = 2.89 \text{ kN}$$

7.2.3   Method of Sections

The method of joints and Maxwell's diagram are useful when the forces in all the members of the truss are required. If, however, the forces in only some of the individual members of a truss are required, the method of sections is more convenient. By using Example 7.1 again and supposing that only the force in member AG is required, then the procedure would be to divide the frame by a line which cuts the member under consideration. The forces acting on the portion of the frame to the left or right of the section are considered, and a free-body diagram for that portion is drawn. Thus, if the truss is divided as shown in Fig. 7.7a, then the free-body diagram for that portion of the truss to the left of the section is as shown in Fig. 7.7b.

Taking moments about joint $DFGHC$ gives

$$F_{AG}(1.5 \sin 60°) + 1.27(1.5) = 0$$

or
$$F_{AG} = \frac{-1.27(1.5)}{1.5(0.866)} = -1.47 \text{ kN}$$

Thus, the force in member AG is 1.47 kN, compression.

7.3 Frames

As stated earlier, a frame is distinguished from a pin-jointed truss by the fact that loads are not necessarily applied at the joints so that at least some, if not all, of the members may be subjected to transverse loads and bending. Thus, the lines of action of forces acting on a member do not necessarily coincide with a straight line passing through the ends of the member.

Example 7.2 Determine the reactions at the pin supports for member DEF in the frame shown in Fig. 7.8. Neglect the weights of the members.

Solution. The various free-body diagrams for the solution of this problem are shown in Fig. 7.9. From Fig. 7.9a it can be seen, by resolving forces in the x direction, that

$$\sum F_x = 0: \qquad\qquad F_{Ah} = 0 \qquad\qquad\qquad\text{(a)}$$

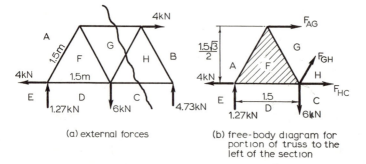

(a) external forces

(b) free-body diagram for portion of truss to the left of the section

Fig. 7.7 Method of sections.

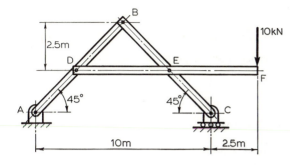

Fig. 7.8 Frame for Example 7.2.

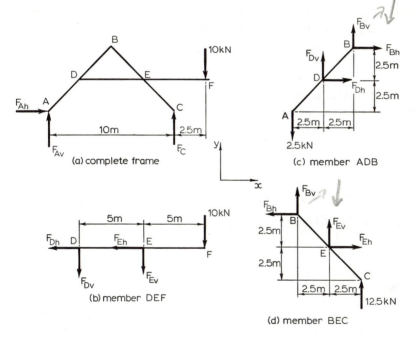

Fig. 7.9 Free-body diagrams for Example 7.2.

Resolving forces in the y direction gives

$$\sum F_y = 0: \qquad\qquad F_{Av} + F_C = 10 \qquad\qquad \text{(b)}$$

Taking moments about point A gives

$$\sum M_A = 0: \qquad\qquad 10F_C = 10(12.5) \qquad\qquad \text{(c)}$$

or

$$\qquad\qquad\qquad F_C = 12.5 \text{ kN} \qquad\qquad \text{(d)}$$

Hence, from Eq. (b),

$$F_{Av} = -2.5 \text{ kN} \tag{e}$$

From Fig. 7.9b, taking moments about point D gives

$$\sum M_D = 0: \qquad 5F_{Ev} = -10(10)$$

or

$$F_{Ev} = -20 \text{ kN} \tag{f}$$

Resolving forces in the y direction gives

$$\sum F_y = 0: \qquad F_{Dv} + F_{Ev} = -10$$

and

$$F_{Dv} = -10 - (-20) = 10 \text{ kN} \tag{g}$$

Now, utilizing the free-body diagram for member ADB (Fig. 7.9c), we can determine F_D by taking moments about point B. Thus,

$$\sum M_B = 0: \qquad 2.5(5) + F_{Dh}(2.5) - F_{Dv}(2.5) = 0$$

or

$$F_{Dh} = F_{Dv} - 5 = 10 - 5 = 5 \text{ kN} \tag{h}$$

Finally, returning to member DEF (Fig. 7.9b) and resolving forces in the horizontal direction, we get

$$\sum F_x = 0: \qquad F_{Eh} = -F_{Dh} = -5 \text{ kN} \tag{i}$$

In summary,

$$F_{Dv} = 10 \text{ kN} \qquad F_{Ev} = -20 \text{ kN}$$
$$F_{Dh} = 5 \text{ kN} \qquad F_{Eh} = -5 \text{ kN}$$

and it is seen that the resultants of these force components do not lie along the axis of the member.

Problems

7.1 For the plane truss shown in Fig. P7.1, determine the force in each member and indicate whether it is a tension or compression member. Neglect the weights of the members.

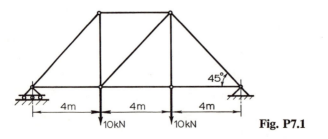

Fig. P7.1

7.2 Figure P7.2 shows a truss used to support a power line. Determine the force in each member of the truss if the force in the cable at *A* is 10 kN.

7.3 For the Howe truss shown in Fig. P7.3, determine the force in member *JD*.

7.4 Determine the forces in members *BE*, *BF*, and *AF* for the railroad crane truss shown in Fig. P7.4. Neglect friction in the pulleys.

7.5 The frame shown in Fig. P7.5 is pin-connected at *B*, *D*, *E*, and *C*. For the given loading, determine the magnitudes of the forces in the pins at *E* and *B*.

7.6 For the pinned arch frame shown in Fig. P7.6, determine the magnitudes of the forces in the pins at *A*, *B*, and *C* for the loading shown.

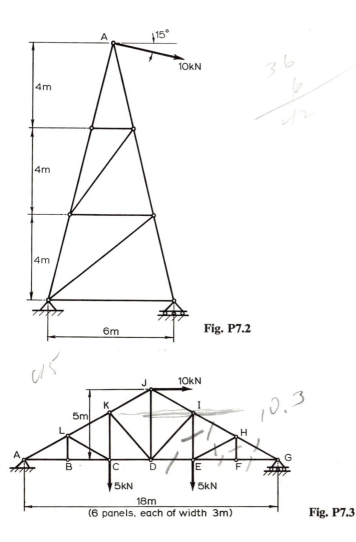

Fig. P7.2

Fig. P7.3

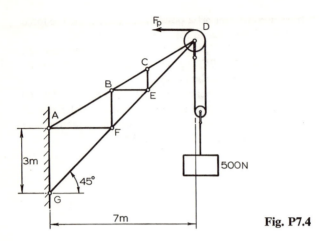

Fig. P7.4

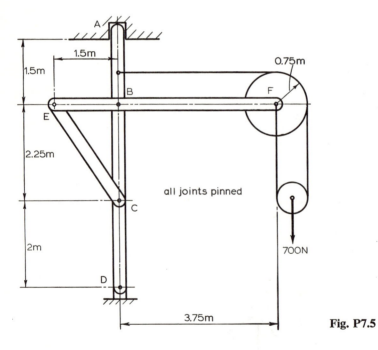

Fig. P7.5

7.7   For the truss shown in Fig. 7.7, determine the force in member *BC*.
7.8   For the pinned frame structure shown in Fig. P7.8, determine the magnitude of the force in the pin at *B*.
7.9   Determine the forces in members *DE* and *GH* for the truss shown in Fig. P7.9.

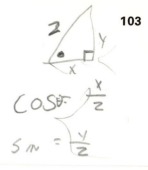

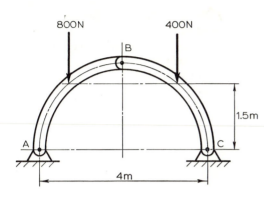

Fig. P7.6

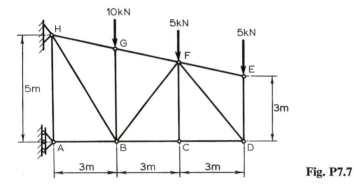

Fig. P7.7

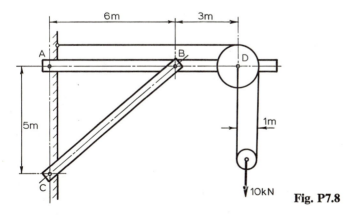

Fig. P7.8

7.10 For the pinned frame structure shown in Fig. P7.10, determine the horizontal and vertical components of the force in the pin at *B*. Neglect the weights of all the members.

7.11   For the symmetrical railway bridge shown in Fig. P7.11, find the forces in
       members *DE*, *CJ*, and *KJ*, and state whether each member is in tension or
       compression. Solve by the method of sections, using only one equation for
       each force where the force appears as the only unknown.

7.12   Find the force in each member of the truss shown in Fig. P7.12.

7.13   Find the force in each member of the truss shown in Fig. P7.13.

7.14   All panels of the truss shown in Fig. P7.14 are 45° right triangles. Determine
       the forces in members *CD*, *DE*, and *CE* in terms of the load F_w.

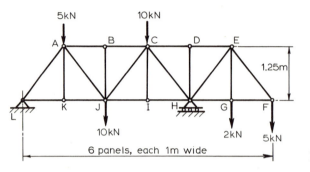

Fig. P7.9

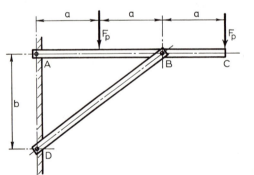

Fig. P7.10

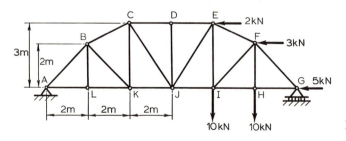

Fig. P7.11

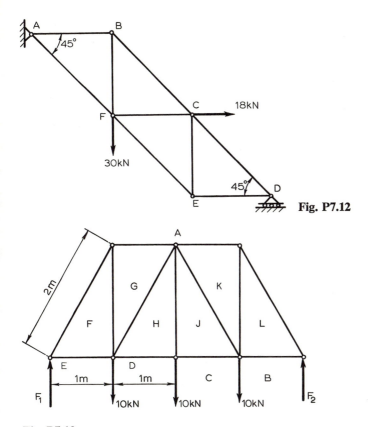

Fig. P7.12

Fig. P7.13

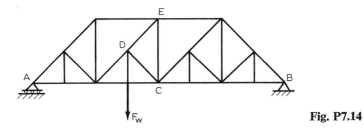

Fig. P7.14

7.15 For the two-member frame pinned at *A, B,* and *C* and loaded by the force and couple shown in Fig. P7.15, determine the resultant forces acting on the pins at *A* and *B.*

7.16 Determine the forces in members *CH, CB,* and *GH* for the truss shown in Fig. P7.16.

7.17 Determine the forces in members *KJ, DE,* and *DI* for the truss shown in Fig. P7.17.

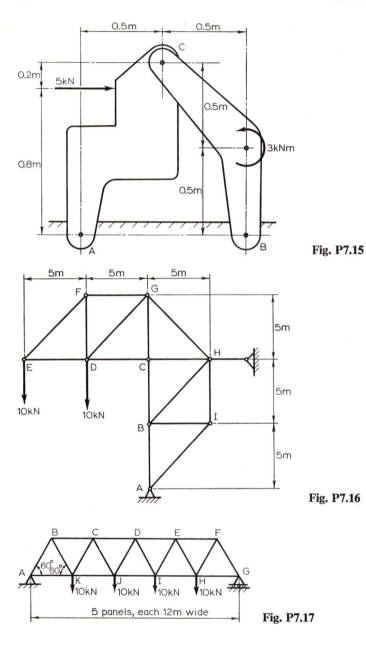

Fig. P7.15

Fig. P7.16

Fig. P7.17

8

Mechanisms

8.1 Introduction

A machine is usually understood to be a mechanical system consisting of various moving parts for doing some kind of useful work. The machines discussed in Chapter 6 consisted of only one or two moving parts and were called simple machines. More elaborate machines often have several moving parts connected together so that one kind of motion is transformed into the motion desired. One example would be the arrangement of connecting rod, crank, and piston in an automobile engine. This arrangement converts the reciprocating motion of the piston to the rotary motion of the crank. Such arrangements of moving parts are called mechanisms.

Mechanisms are similar to trusses and frames except that the joints are truly pinned and the individual members are expected to move relative to one another. The way that the members of a mechanism move falls within the subject of kinematics, which will be studied later. In the present chapter we shall simply examine the forces transmitted from one member to another in a mechanism when it is fixed in one particular position. Under these conditions, the forces transmitted from one member to another can be determined by the conditions for static equilibrium.

8.2 Examples of Mechanisms

Example 8.1   Figure 8.1a shows a simple mechanism for gripping an object. Determine the gripping force F_g if the force applied to each handle is F_p.

Solution.   To find the gripping force F_g we can isolate one of the two identical

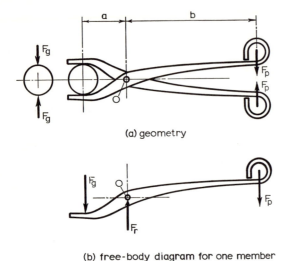

(a) geometry

(b) free-body diagram for one member

Fig. 8.1 A simple mechanism for gripping.

members of the mechanism and draw the free-body diagram shown in Fig. 8.1b. Since the two forces F_p and F_g are vertical, the reaction F_r at the pin joint must also be vertical.

Taking moments about the pin joint, we get

$$\sum M_O = 0: \qquad\qquad F_g a - F_p b = 0$$

or

$$F_g = F_p \frac{b}{a}$$

Example 8.2 An example of another simple mechanism, known as a four-bar linkage, is shown in one particular position in Fig. 8.2. If the torque (moment) in the drive shaft at D is M_D, determine the torque M_A on the shaft at A to maintain static equilibrium. Neglect the masses of the members and friction in the joints.

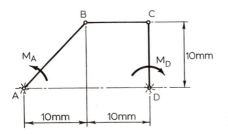

Fig. 8.2 Four-bar linkage.

Solution. A free-body diagram can be drawn for each of the links or members in this mechanism (Fig. 8.3). However, it should first be realized that since the link BC (Fig. 8.3a) has only two direct forces acting on it, the lines of action of these forces must pass through the joints B and C. Thus, resolving horizontally, we get

$$\sum F_x = 0: \qquad\qquad F_B = F_C \qquad\qquad\qquad (a)$$

Now, referring to the free-body diagram for the link AB (Fig. 8.3b) and taking moments about point A, we get

$$\sum M_A = 0: \qquad\qquad M_A = F_B(10) \qquad\qquad\qquad (b)$$

Finally, using the free-body diagram for the link CD (Fig. 8.3c) and taking moments about D, we get

$$\sum M_D = 0: \qquad\qquad M_D = F_C(10) \qquad\qquad\qquad (c)$$

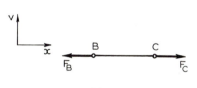

(a) link BC

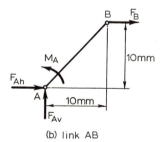

(b) link AB

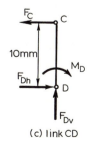

(c) link CD

Fig. 8.3 Free-body diagrams for four-bar linkage.

Combining Eqs. (a), (b), and (c), we get

$$M_A = M_D$$

Example 8.3   Figure 8.4 shows the mechanism used in internal combustion engines; it is known as a slider-crank or simple-engine mechanism. It is used to convert the translational motion of the piston C into rotational motion of the crank AB. If the instantaneous force on the piston due to the pressure in the cylinder is F_p, determine the moment M_A on the shaft at A in order to maintain static equilibrium. Neglect friction and the masses of the linkages and the piston.

Solution.   If l is the length of the connecting rod and r the radius or "throw" of the crank, we can determine the moment or torque M_A on the crankshaft when the crank is positioned at an angle θ as shown. From the free-body diagram for the connecting rod BC (Fig. 8.5a) and resolving forces parallel to BC, we get

$$\sum F_x = 0: \qquad\qquad F_B = F_p \cos \phi + F_C \sin \phi \qquad\qquad \text{(a)}$$

Resolving forces perpendicular to BC, we get

$$\sum F_y = 0: \qquad\qquad F_p \sin \phi = F_C \cos \phi \qquad\qquad \text{(b)}$$

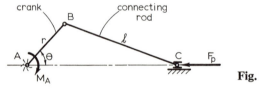

Fig. 8.4   Slider-crank mechanism.

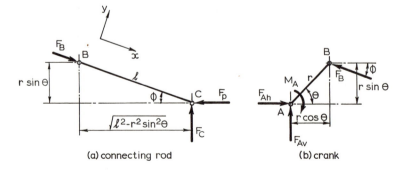

(a) connecting rod   (b) crank

Fig. 8.5   Free-body diagrams for slider-crank mechanism.

Eliminating F_C from Eqs. (a) and (b), we get

$$F_B = F_p\left(\cos \phi + \frac{\sin^2 \phi}{\cos \phi}\right)$$

or
$$F_B = \frac{F_p}{\cos \phi} \tag{c}$$

From the free-body diagram for the crank AB (Fig. 8.5b) and taking moments about A, we get

$$\sum M_A = 0: \qquad M_A = F_B \cos \phi(r \sin \theta) + F_B \sin \phi(r \cos \theta)$$
$$= F_B r(\cos \phi \sin \theta + \sin \phi \cos \theta) \tag{d}$$

Substitution for F_B from Eq. (c) into Eq. (d) gives

$$M_A = F_p r(\sin \theta + \cos \theta \tan \phi) \tag{e}$$

Now, by geometry,

$$\tan \phi = \frac{r \sin \theta}{\sqrt{l^2 - r^2 \sin^2 \theta}} \tag{f}$$

and substitution of Eq. (f) into Eq. (e) gives

$$M_A = F_p r \sin \theta\left(1 + \frac{r \cos \theta}{\sqrt{l^2 - r^2 \sin^2 \theta}}\right) \tag{g}$$

Problems

8.1   Figure P8.1 shows a Geneva drive, a mechanism that converts a constant rotary motion into rotary indexing motion. If the torque on the drive shaft A is M_A, obtain an expression for the instantaneous torque M_B on the driven shaft at B.

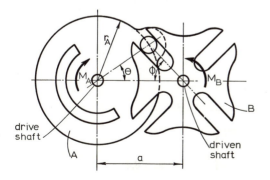

Fig. P8.1

8.2 The toggle mechanism shown in Fig. P8.2 is used in rock crushers, presses, riveting machines, and clutches, where a large force is required to act through a small distance. If the input force F_p for the position shown is 400 N, determine the horizontal force exerted on the block D. Neglect friction.

8.3 For the tongs shown in Fig. P8.3, determine the magnitudes of forces in the pins at A and B and the magnitude of the resultant force on the block at C.

8.4 Determine the force exerted by the cutter on each side of the bolt if the force applied to each handle is 250 N (Fig. P8.4).

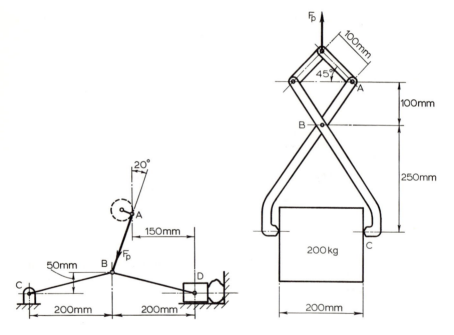

Fig. P8.2   **Fig. P8.3**

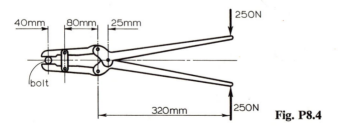

Fig. P8.4

8.5 For the loader mechanism shown in Fig. P8.5, determine the force in the hydraulic actuator DG and the reaction at the pin connection E to maintain static equilibrium. The bucket is supported by two identical mechanisms side by side. One-half of the total load F_w of 40 kN is applied to each mechanism.

8.6 A mechanism is shown in Fig. P8.6 at a particular instant. If the torque in the drive shaft at D is M_D, determine the torque M_A in the shaft at A required to maintain static equilibrium. Neglect friction.

8.7 The hydraulic jack shown in Fig. P8.7 is used to raise a 3-kN load. Determine the magnitude of the force in the pin at A and the force in the hydraulic actuator.

8.8 For the press mechanism shown in Fig. P8.8, determine the force acting on the plate C if a 400-N force is applied normal to the press handle. In addition, determine the magnitudes of the forces in the pins at A and B.

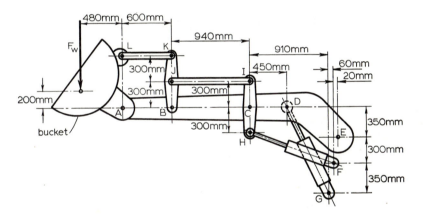

Fig. P8.5

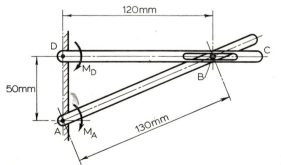

Fig. P8.6

8.9   Figure P8.9 shows a device to limit the tension in a cable and consists of the
      toggle-operated jaws which shear the pin A when the cable tension exceeds
      a predetermined value. If the pin will resist up to 12 kN of shear force before
      it shears, determine the maximum tension F_t that the device will permit in the
      cable.

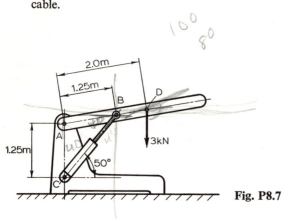

Fig. P8.7

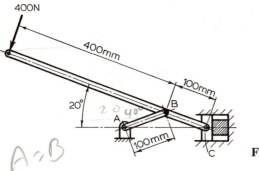

Fig. P8.8

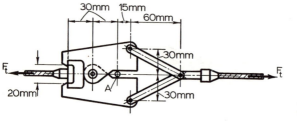

Fig. P8.9

9

Beams

9.1 Introduction

A beam is a structural member subjected primarily to bending. Most beams are long prismatic bars with the loads usually applied in a direction perpendicular to the longitudinal axis of the beam. A straight beam that is resting on two supports is called a simply supported beam (Fig. 9.1a). A straight beam that is rigidly held at one end is called a cantilever beam (Fig. 9.1b).

Because beams play such an important role in the design of engineering structures, an understanding of the load-carrying capabilities of beams is

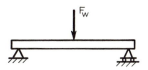

(a) simply supported beam

(b) cantilever beam   **Fig. 9.1** Straight beams.

important. A beam analysis first consists of the determination of all the external forces including loads and reactions acting on the beam, using static equilibrium conditions, and then a study of the effect of these forces on the bending deflections of the beam and internal stresses in the beam. The latter study requires a knowledge of the strength characteristics of the material from which the beam is made and is usually treated in texts dealing with the mechanics of deformable bodies or strengths of materials. The former study involves an application of the principles of statics and will be dealt with in this chapter.

9.2   Bending Moments and Shear Forces

Figure 9.2a shows the arrangement of a straight, horizontal beam simply supported at A and B and subjected to a system of vertical loads F_1, F_2, and F_3 at various points along the center plane of the beam. The reactions at the supports are F_A and F_B. If the beam were cut through at any section k-k (Fig. 9.2a), then a vertical force F_s and a couple or moment M would have to be applied at the cut section in order to maintain static equilibrium of either portion of the beam (Fig. 9.2b). The vertical force F_s is called the shear force, while the couple or moment M is called the bending moment. Since the portion of the beam shown in Fig. 9.2b must be in static equilibrium, the shear force and bending moment may be defined as follows:

The *shear force* F_s at a section is the algebraic sum of the forces acting to the right or left of the section, resolved parallel to the plane of the section if necessary.

The *bending moment* M at a section is the algebraic sum of the moments of

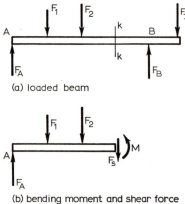

(a) loaded beam

(b) bending moment and shear force
   at section k-k.   **Fig. 9.2**   Simply supported beam.

all the forces acting to the right or left of the section, the moments being calculated with reference to an axis in the plane of the section.

9.3 Sign Convention

It is generally convenient to adopt a sign convention for the positive directions of F_s and M, and Fig. 9.3 shows the convention usually employed. Positive shear is where the resultant force acting on the portion of the beam to the right of the section considered acts downward, while a positive bending moment causes the beam to bend such that its top surface is concave.

9.4 Shear-Force and Bending-Moment Diagrams

Generally the values of the shear force and the bending moment vary from section to section along the beam. Therefore, it is usual to construct shear-force and bending-moment diagrams so that the variation in these quantities can be observed and the design of the beam facilitated.

Example 9.1 For the configuration shown in Fig. 9.4a, draw the corresponding shear-force and bending-moment diagrams.

Solution. To determine the reaction forces we use the free-body diagram shown in Fig. 9.4b. Taking moments about A gives

$$\sum M_A = 0: \qquad 4(5) + 6(10) - F_D(20) + 2(25) = 0$$

or

$$F_D = \frac{130}{20} = 6.5 \text{ kN} \qquad\qquad (a)$$

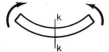

positive bending

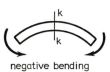

negative bending

positive shear

negative shear

Fig. 9.3 Sign convention for shear force and bending moment.

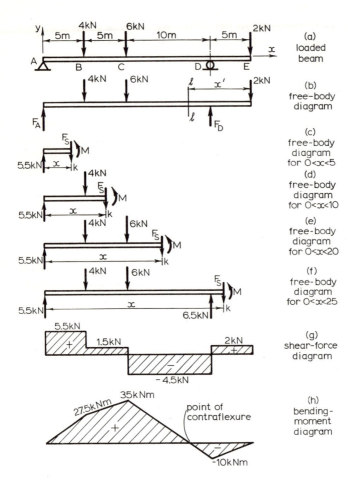

Fig. 9.4 Diagrams for Example 9.1.

Resolving vertically, we get

$$\sum F_y = 0: \qquad F_A + F_D - 4 - 6 - 2 = 0$$

$$F_A = -F_D + 12 = -6.5 + 12$$

$$F_A = 5.5 \text{ kN} \tag{b}$$

To construct the shear-force and bending-moment diagrams it is necessary to obtain the values of the shear force and the bending moment at various sections along the beam. The free-body diagram for portion AB of the beam ($0 < x < 5$) is shown in Fig. 9.4c. Thus, for $0 < x < 5$,

$$\sum F_y = 0: \qquad\qquad F_s = 5.5 \text{ kN} \qquad\qquad\qquad \text{(c)}$$
$$\sum M_k = 0: \qquad\qquad M = 5.5x \qquad\qquad\qquad\qquad \text{(d)}$$

which means that, in this interval, F_s is positive and constant (Fig. 9.4g), while M is positive and increases linearly from zero at A (Fig. 9.4h).

Portion BC of the beam is now analyzed using the free-body diagram shown in Fig. 9.4d. Thus, for $5 < x < 10$,

$$\sum F_y = 0: \qquad F_s = 5.5 - 4 = 1.5 \text{ kN} \qquad\qquad\qquad \text{(e)}$$
$$\sum M_k = 0: \qquad M = 5.5x - 4(x - 5) = 1.5x + 20 \qquad\qquad \text{(f)}$$

A similar procedure is followed for portions CD and DE. Thus, for $10 < x < 20$,

$$\sum F_y = 0: \qquad F_s = 5.5 - 4 - 6 = -4.5 \text{ kN} \qquad\qquad \text{(g)}$$
$$\sum M_k = 0: \qquad M = 5.5x - 4(x - 5) - 6(x - 10)$$
$$= -4.5x + 80 \qquad \text{(h)}$$

and for $20 < x < 25$,

$$\sum F_y = 0: \qquad F_s = 5.5 - 4 - 6 + 6.5 = 2 \text{ kN} \qquad\qquad \text{(i)}$$
$$\sum M_k = 0: \qquad M = 5.5x - 4(x - 5) - 6(x - 10) + 6.5(x - 20)$$
$$= 2x - 50 \qquad \text{(j)}$$

The completed shear-force and bending-moment diagrams are shown in Figs. 9.4g and 9.4h, respectively. It will be seen that a point of contraflexure, where the bending moment changes sign, occurs between C and D; its position can be found as follows: Considering section l-l distance x' from the right-hand end of the beam, we get

$$M_l = 2x' - 6.5(x' - 5) = 0$$
$$2x' = 6.5(x' - 5)$$
$$x' = \frac{32.5}{4.5} = 7.22 \text{ m}$$

9.4.1 Standard Cases

Figure 9.5 shows the shear-force and bending-moment diagrams for simply supported and cantilever beams with various standard types of loading.

9.5 Relations Among Load, Shear Force, and Bending Moment

A horizontal beam simply supported at each end and carrying a general form of distributed loading is shown in Fig. 9.6a. We shall consider the equilibrium

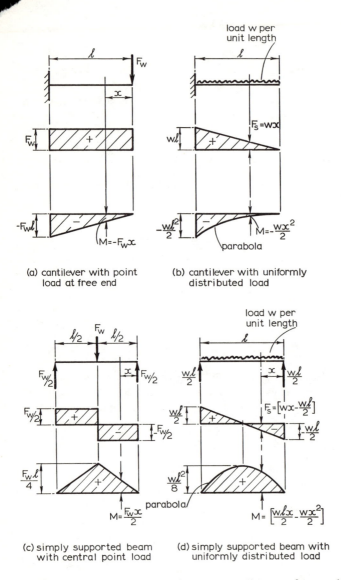

Fig. 9.5 Shear-force and bending-moment diagrams for standard cases.

of an element of the beam of length dx and at a distance x from the left-hand end. Figure 9.6b shows the free-body diagram for this element. The length dx is considered so small that variations in loading along this length can be ignored. Resolving vertically, we get

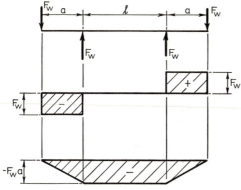

(e) simply supported beam with equal overhanging point loads

Fig. 9.5   Continued.

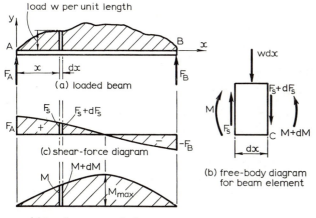

Fig. 9.6   Beam with distributed load.

$$\sum F_y = 0: \qquad\qquad F_s - (F_s + dF_s) - w\,dx = 0$$

or

$$w = -\frac{dF_s}{dx} \qquad\qquad (9.1)$$

Integrating, we get

$$\int_{F_{s1}}^{F_{s2}} dF_s = -\int_{x_1}^{x_2} w\,dx$$

or

$$F_{s2} - F_{s1} = -\int_{x_1}^{x_2} w\,dx \qquad\qquad (9.2)$$

Therefore, the change in shear force between any two sections of beam is equal to the negative of the area of the loading diagram between those sections. Taking moments about C, we get

$$\sum M_C = 0: \qquad M - (M + dM) - w\, dx\left(\frac{dx}{2}\right) + F_s\, dx = 0 \qquad (9.3)$$

Neglecting second-order terms, we get

$$\frac{dM}{dx} = F_s \qquad (9.4)$$

and between any two sections,

$$\int_{M_1}^{M_2} dM = \int_{x_1}^{x_2} F_s\, dx$$

or $\qquad\qquad\qquad M_2 - M_1 = \int_{x_1}^{x_2} F_s\, dx \qquad (9.5)$

Therefore, the change in bending moment between any two sections of the beam is equal to the area of the shear-force diagram between those sections.

From Eq. (9.4) it will be noted that the bending moment is a maximum or minimum when the shear force is zero or, more correctly, when F_s is changing sign.

Example 9.2 For the configuration shown in Fig. 9.7a, draw the shear-force and bending-moment diagrams.

Solution. Taking moments about B gives

$$\sum M_B = 0: \qquad 10F_D + 8(5) - 1.5(2)(13) - 10(15) = 0$$
$$F_D = 14.9 \text{ kN} \qquad (a)$$

Resolving forces in the y direction gives

$$\sum F_y = 0: \qquad\qquad F_B = 8 + 10 + 2(13) - 14.9$$
$$= 29.1 \text{ kN} \qquad (b)$$

The shear force at point A is -8 kN; thus, the shear-force diagram (Fig. 9.7b) starts at A with a value of -8 kN. Along section AB of the beam, the loading is 2 kN/m so that the change in shear force from A to B is given by

$$F_{sB} - F_{sA} = -(\text{area of the loading diagram})$$
$$= -(2)(5) = -10 \text{ kN} \qquad (c)$$

Hence, the negative shear force increases linearly from a value of -8 kN at A to -18 kN at B where there is a step change in the shear force of $+29.1$ kN, i.e., from -18 kN to 11.1 kN. Along BC, the area of the loading diagram is 16 kN so that the shear force decreases linearly from 11.1 kN at B to -4.9 kN at C. Since no loads are applied within the section CD, the shear force re-

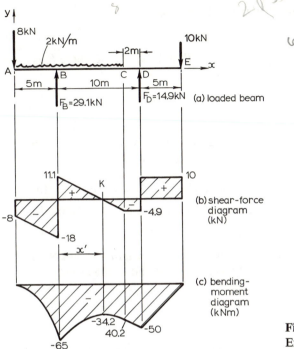

(a) loaded beam

(b) shear-force diagram (kN)

(c) bending-moment diagram (kNm)

Fig. 9.7 Diagrams for Example 9.2.

mains constant from C to D. At D there is a step change in the shear force of $+14.9$ kN, i.e., from -4.9 kN to 10 kN. The shear force remains constant at a value of 10 kN from D to E.

Figure 9.7c shows the corresponding bending-moment diagram. The bending moment is zero at A since no moments are applied to the left of this section. From Eq. (9.5), the change in bending moment from A to B is equal to the area under the shear-force diagram or

$$M_B - M_A = (-8)(5) + \tfrac{1}{2}(-5)(10) = -65 \text{ kN} \cdot \text{m} \qquad \text{(d)}$$

Thus,

$$M_B = -65 \text{ kN} \cdot \text{m} \qquad \text{(e)}$$

When the shear force changes sign, a maximum or minimum occurs in the bending moment. To determine the section K in the beam where this occurs we can write the expression for F_s and then equate F_s to zero. Thus, for section BC,

$$F_s = 11.1 - 2x'$$

when x' is measured from B to the section. Thus,

$$x' = 5.55 \text{ m} \tag{f}$$

Now the change in bending moment from B to the section K where F_s is zero is given by

$$M_K - M_B = \tfrac{1}{2}(11.1)(5.55) = 30.8 \text{ kN} \cdot \text{m}$$

Hence,

$$M_K = -65 + 30.8 = -34.2 \text{ kN} \cdot \text{m} \tag{g}$$

Similarly,

$$M_C = -34.2 + \tfrac{1}{2}(-4.9)(2.45) = -40.2 \text{ kN} \cdot \text{m} \tag{h}$$

From C to D, the area under the shear-force diagram is -9.8 kN·m. Hence,

$$M_D = -40.2 - 9.8 = -50 \text{ kN} \cdot \text{m} \tag{i}$$

Since from D to E the area under the shear-force diagram is 50 kN·m, the bending moment decreases linearly to zero at the end of the beam.

Problems

9.1 For the configurations shown in Fig. P9.1, draw the shear-force and bending-moment diagrams.

9.2 A load of w N/m distributed uniformly over a length l is transferred to a soil foundation by means of a timber beam of length $2l$ (Fig. P9.2). If the reaction between the soil and the beam is uniformly distributed over the length of the beam, sketch the shear-force and bending-moment diagrams. Neglect the weight of the beam.

9.3 For the beam and loading shown in Fig. P9.3, draw the shear-force and bending-moment diagrams.

9.4 A cantilever beam is loaded as shown in Fig. P9.4. Draw the shear-force and bending-moment diagrams for this beam. What is the value of the maximum shear force, and at what section does this occur? What is the value of the maximum bending moment, and at what section does this occur? Neglect the weight of the beam.

9.5 Draw the shear-force and bending-moment diagrams for the diving board shown in Fig. P9.5 when the 80-kg diver is standing at the end of the board.

9.6 For the simply supported beam shown in Fig. P9.6, sketch the shear-force and bending-moment diagrams. Show all the principal values including the magnitude and position of the maximum bending moment within the section AB.

9.7 Referring to Fig. P9.7, derive expressions for the bending moments in the center of the beam and at the load F_w.

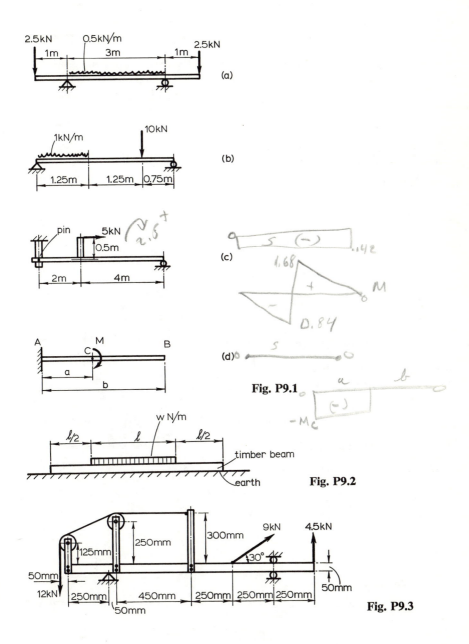

Fig. P9.1

Fig. P9.2

Fig. P9.3

9.8   For the simply supported beam shown in Fig. P9.8, sketch the bending-moment and shear-force diagrams, and indicate all the principal values.

9.9   For the arrangement shown in Fig. P9.9, draw the shear-force and bending-moment diagrams, showing the principal values, and obtain an expression for the maximum bending moment.

9.10 For the simply supported beam shown in Fig. P9.10, sketch the shear-force and bending-moment diagrams, showing the principal values, and determine the position and magnitude of the maximum bending moment.

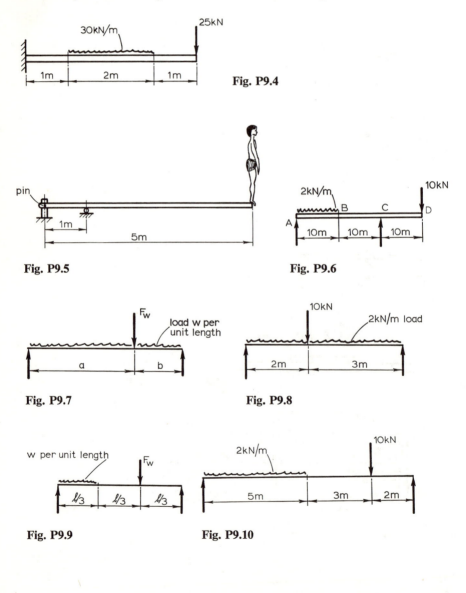

Fig. P9.4

Fig. P9.5

Fig. P9.6

Fig. P9.7

Fig. P9.8

Fig. P9.9

Fig. P9.10

10

Common Problems Involving Friction

10.1 Introduction

Friction arises in all machinery where contacting parts move relative to each other. In many cases, it lowers the efficiency of the machine and causes wear, and steps are taken to reduce friction as much as possible by lubrication. However, there are many situations where friction is necessary for operation of the machine, for example, in belt drives, clutches, holding screws, etc. In this chapter, analyses are presented of situations both where friction is necessary and where friction is undesirable. In all cases, the expressions developed apply equally to static and dynamic frictional conditions. Hence, in the analyses, the coefficient of friction μ will be employed, but it should be remembered that this can be either the coefficient of static friction μ_s or the coefficient of dynamic friction μ_d depending on the circumstances.

10.2 Belt Friction

When a rope or belt is coiled around a shaft or pulley (Fig. 10.1) the tendency for the rope or belt to slip depends substantially on the angle subtended by the arc of contact. For example, a rope coiled around a ship's capstan (Fig. 10.2) is able to provide so much resistance to sliding that the ship can be pulled toward the dock by rotating the capstan. This frictional resistance is known as *belt friction*, a phenomenon which also allows the transmission of large quantities of power through belt and rope drives.

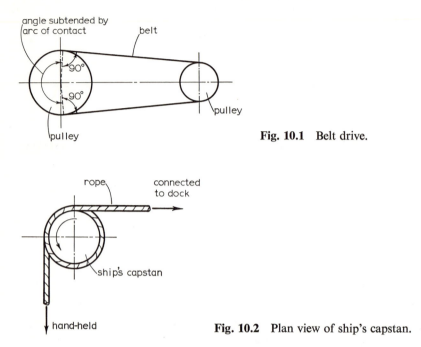

Fig. 10.1 Belt drive.

Fig. 10.2 Plan view of ship's capstan.

Figure 10.3 shows a belt coiled around a pulley together with a free-body diagram for an element of the belt. The belt is in contact with the pulley over an arc which subtends an angle θ_w at the center of the pulley. Tensions F_{t1} and F_{t2} are applied to the ends of the belt, F_{t1} being greater than F_{t2}. The relative magnitudes of F_{t1} and F_{t2} are such that the belt is just about to slip. The forces acting on the small element of the belt are the normal reaction dF_n between the belt and the pulley, the limiting frictional resistance $\mu\, dF_n$, and the tensions F_t and $F_t + dF_t$ at each end of the element. Resolving vertically, we get

$$\sum F_y = 0: \qquad dF_n - F_t \sin\frac{d\theta}{2} - (F_t + dF_t)\sin\frac{d\theta}{2} = 0$$

or when $d\theta$ is small then $\sin(d\theta/2)$ is approximately equal to $d\theta/2$, and upon neglecting higher-order terms,

$$dF_n = F_t\, d\theta \qquad\qquad (10.1)$$

Resolving horizontally, we get

$$\sum F_x = 0: \qquad F_t \cos\frac{d\theta}{2} + \mu\, dF_n - (F_t + dF_t)\cos\frac{d\theta}{2} = 0$$

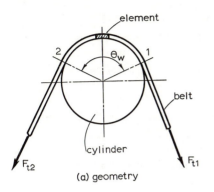

(a) geometry

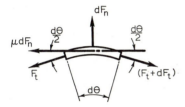

(b) free-body diagram for element   **Fig. 10.3**   Belt coiled around a cylinder.

or when $d\theta$ is small then $\cos(d\theta/2)$ is approximately equal to unity, and

$$\mu \, dF_n = dF_t \tag{10.2}$$

Eliminating dF_n from Eqs. (10.1) and (10.2) gives

$$\frac{dF_t}{F_t} = \mu \, d\theta \tag{10.3}$$

Equation (10.3) is a first-order differential equation and is integrated between points 1 and 2 where the belt loses contact with the pulley. Thus,

$$\int_{F_{t2}}^{F_{t1}} \frac{dF_t}{F_t} = \mu \int_0^{\theta_w} d\theta$$

and

$$\ln\left(\frac{F_{t1}}{F_{t2}}\right) = \mu\theta_w$$

or

$$\frac{F_{t1}}{F_{t2}} = e^{\mu\theta_w} \tag{10.4}$$

It can be seen that the ratio F_{t1}/F_{t2} increases exponentially with increasing values of θ_w and explains why such large values of frictional resistance can be achieved. To increase this tension ratio in belt drives, an idler wheel (Fig. 10.4) is sometimes used to increase θ_w. The angle θ_w is called the angle of wrap. Since F_{t1} is greater than F_{t2}, F_{t1} is called the tight-side belt tension, while F_{t2} is called the slack-side belt tension.

An alternative method of increasing the tension ratio is to replace the flat belt and pulley (Fig. 10.3) by a V-belt and matching grooved pulley, as shown in Fig. 10.5a. A free-body diagram for an element of a V-belt is shown in Fig. 10.5b. If the relative magnitudes of F_{t1} and F_{t2} are such that the belt is just about to slip, then resolving vertically, we get

$$\sum F_y = 0: \quad 2dF_n \sin \beta - F_t \sin \frac{d\theta}{2} - (F_t + dF_t) \sin \frac{d\theta}{2} = 0$$

By replacing $\sin(d\theta/2)$ by $d\theta/2$ and neglecting higher-order terms, this equation becomes

$$2dF_n \sin \beta = F_t \, d\theta \tag{10.5}$$

Resolving horizontally, we get

$$\sum F_x = 0: \quad (F_t + dF_t) \cos \frac{d\theta}{2} - F_t \cos \frac{d\theta}{2} - 2\mu \, dF_n = 0$$

or when $\cos(d\theta/2)$ is replaced by unity,

$$dF_t = 2\mu \, dF_n \tag{10.6}$$

Eliminating dF_n from Eqs. (10.5) and (10.6) gives

$$\frac{dF_t}{F_t} = \frac{\mu}{\sin \beta} \, d\theta = \mu_e \, d\theta \tag{10.7}$$

where

$$\mu_e = \frac{\mu}{\sin \beta} \tag{10.8}$$

and can be regarded as the equivalent coefficient of friction. Hence, for a V-belt,

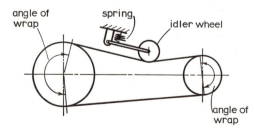

angle of wrap   spring   idler wheel

angle of wrap

Fig. 10.4 Idler wheel used to increase angle of wrap in belt drives.

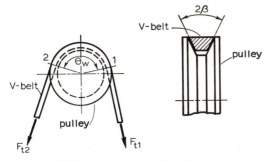

(a) V-belt and grooved pulley

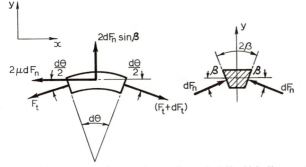

(b) free-body diagram for an element of the V-belt   **Fig. 10.5** V-belt.

$$\frac{F_{t1}}{F_{t2}} = e^{\mu_e \theta_w} \tag{10.9}$$

If, in a particular situation, the angle 2β were $35°$, then from Eq. (10.8) μ_e would be equal to 3.3μ. Thus, the use of a $35°$ V-velt would be equivalent to increasing the coefficient of friction of a flat belt of the same material by a factor of 3.3.

Example 10.1   A rope is coiled $3\frac{1}{4}$ turns around an 0.5-m-diameter ship's capstan (Fig. 10.6). One end of the rope is secured to the dock and the other end held by a sailor. With what force must the sailor pull his end of the rope to withstand a tension of 10 kN in the section of the rope between the ship and the dock? The coefficient of friction between the rope and the capstan is 0.2.

Solution.   In this problem, the angle of wrap is

$$\theta_w = 3.25(360°) = 1170° = 20.42 \text{ rad}$$

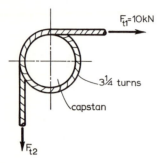

Fig. 10.6   Diagram for Example 10.1.

From Eq. (10.4),

$$\frac{F_{t1}}{F_{t2}} = e^{\mu\theta_w} = e^{0\cdot2(20\cdot42)} = 59.39$$

Therefore, the force applied by the sailor is given by

$$F_{t2} = \frac{F_{t1}}{59.39} = \frac{10 \times 10^3}{59.39} = 168 \text{ N}$$

10.3   Disk Friction

Disk friction arises when two circular (or annular) surfaces are placed in contact and rotate relative to each other. In practice, this situation occurs in certain clutches where the frictional resistance should usually be as high as possible and in certain bearings where the frictional resistance should usually be as low as possible.

Figure 10.7 shows the arrangement diagrammatically where the normal force on the interface is F_n. We shall now consider an elemental ring of radius r and width dr on which the pressure (normal load per unit area) is p and where the coefficient of friction is μ.

The total normal load on the elemental ring will be $2\pi pr\, dr$, and the frictional resistance due to this load will be $2\pi\mu pr\, dr$. Thus, the frictional moment (or torque) dM generated over the area of the ring will be given by

$$dM = 2\pi\mu pr^2\, dr$$

Hence, the total frictional moment (or torque) will be

$$M = 2\pi\mu \int_{r_1}^{r_2} pr^2\, dr \qquad (10.10)$$

Again, the normal load on the elemental ring is

$$dF_n = 2\pi pr\, dr$$

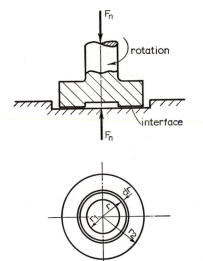

Fig. 10.7 Disk friction.

and the total normal load will be

$$F_n = 2\pi \int_{r_1}^{r_2} pr \, dr \qquad (10.11)$$

To solve Eqs. (10.10) and (10.11) it is necessary to know the relation (if any) between the radius r and the normal load per unit area (pressure) at that radius.

10.3.1 Constant-Pressure Theory

When the analysis above is employed for lubricated bearings it is often reasonable to assume that, because of the supporting oil film, the pressure p is constant and independent of radius. In this case, Eq. (10.10) becomes

$$M = 2\pi\mu p \int_{r_1}^{r_2} r^2 \, dr$$
$$= \tfrac{2}{3}\pi\mu p(r_2^3 - r_1^3) \qquad (10.12)$$

and Eq. (10.11) becomes

$$F_n = 2\pi \int_{r_1}^{r_2} r \, dr$$
$$= \pi p(r_2^2 - r_2^2) \qquad (10.13)$$

Eliminating p from Eqs. (10.12) and (10.13), we get

$$M = \frac{2}{3}\mu F_n \left(\frac{r_2^3 - r_1^3}{r_2^2 - r_1^2} \right) \qquad (10.14)$$

10.3.2   Constant-Wear Theory

The constant-pressure theory cannot be applied to friction clutches or brakes (Fig. 10.8) such as those in an automobile. Friction clutches or brakes are usually designed to permit the disengagement of the coupled elements during rotation. Clearly, clutches or brakes are unlubricated, and the two rubbing surfaces will be in direct contact and will gradually wear due to repeated use.

In general, the wear between two rubbing surfaces is proportional to the normal load per unit area p and proportional to the relative velocity of sliding. Since, in disk friction, the velocity of sliding at any point is proportional to the radius r of that point, the wear will be proportional to both p and r, or, in other words, proportional to pr. Clearly, the wear on an annular clutch face must be constant over the contact area, and therefore pr must be constant. Thus, we may assume that

$$pr = k \tag{10.15}$$

where k is a constant. Substitution of Eq. (10.15) into Eq. (10.11) gives

$$M = 2\pi\mu k \int_{r_1}^{r_2} r\, dr$$
$$= \pi\mu k(r_2^2 - r_1^2) \tag{10.16}$$

and substitution of Eq. (10.15) into Eq. (10.11) gives

$$F_n = 2\pi k \int_{r_1}^{r_2} dr$$
$$= 2\pi k(r_2 - r_1) \tag{10.17}$$

Elimination of k between Eqs. (10.16) and (10.17) gives

$$M = \tfrac{1}{2}\mu F_n(r_2 + r_1) \tag{10.18}$$

Example 10.2   The shaft shown in Fig. 10.9 is subjected to an axial load of 10 kN. The coefficient of dynamic friction between the collar and the support is 0.15. Assuming that the pressure is uniform over the annular contact, determine the moment required to rotate the shaft.

Solution.   From Fig. 10.9, r_2 is 50 mm and r_1 is 25 mm. By using the constant-pressure theory (Eq. 10.14), the moment or torque is given by

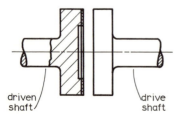

driven
shaft

drive
shaft   **Fig. 10.8**   Friction clutch.

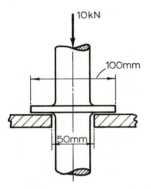

Fig. 10.9 Diagram for Example 10.2.

$$M = \frac{2}{3}\mu F_n\left(\frac{r_2^3 - r_1^3}{r_2^2 - r_1^2}\right)$$
$$= \frac{2}{3}(0.1)(10 \times 10^3)\left[\frac{(0.05)^3 - (0.025)^3}{(0.05)^2 - (0.025)^2}\right] = 39 \text{ N}\cdot\text{m}$$

Example 10.3 Determine the maximum torque (moment) that can be transmitted by the clutch shown in Fig. 10.10. Assume that the coefficient of static friction is 0.4, that the maximum average normal pressure that can be applied is 0.9 kN/m², and that the clutch surface wears uniformly.

Solution. From Fig. 10.10, r_2 is 0.3 m and r_1 is 0.125 m. Thus, the total axial force F_n on the clutch is given by

$$F_n = \pi p_{av}(r_2^2 - r_1^2)$$

where p_{av} is the average pressure and $\pi(r_2^2 - r_1^2)$ is the area of the clutch surface. Hence,

$$F_n = \pi(900)[(0.3)^2 - (0.125)^2] = 210.3 \text{ N}$$

The maximum torque is given by Eq. (10.18). Thus,

$$M = \frac{1}{2}\mu F_n(r_2 + r_1)$$
$$= \frac{1}{2}(0.4)(210.3)(0.3 + 0.125) = 17.88 \text{ N}\cdot\text{m}$$

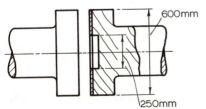

Fig. 10.10 Diagram for Example 10.3.

10.4   Wedge

A wedge is a device used to produce small adjustments in the position of a body. Figure 10.11a shows a wedge A being pushed under a block B of weight F_w. In this problem, relative motion is occurring among three contacting pairs of surfaces [numbered (1), (2), and (3)] simultaneously. At interface (1) the wedge A is moving to the left relative to the ground, at interface (2) the block B is moving to the right relative to the wedge, and at interface (3) the block B is moving upward relative to the wall. For the solution of this problem, it is necessary to consider the free-body diagrams for the wedge only and the wedge and block together and to show the appropriate directions of the

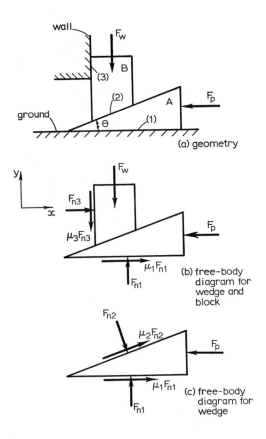

(a) geometry

(b) free-body diagram for wedge and block

(c) free-body diagram for wedge

Fig. 10.11   Wedge and block.

frictional forces. If μ_1, μ_2, and μ_3 are the coefficients of friction for the interfaces (1), (2), and (3), respectively, then for the wedge only (Fig. 10.11c),

$$\sum F_x = 0: \qquad F_p = \mu_1 F_{n1} + F_{n2} \sin \theta + \mu_2 F_{n2} \cos \theta$$

$$= \mu_1 F_{n1} + F_{n2}(\sin \theta + \mu_2 \cos \theta) \tag{10.19}$$

Resolving vertically, we get

$$\sum F_y = 0: \qquad F_{n1} + \mu_2 F_{n2} \sin \theta = F_{n2} \cos \theta \tag{10.20}$$

Eliminating F_{n2} from Eqs. (10.19) and (10.20), we get

$$F_p = F_{n1}\left(\mu_1 + \frac{\sin \theta + \mu_2 \cos \theta}{\cos \theta - \mu_2 \sin \theta}\right) \tag{10.21}$$

From the free-body diagram for the wedge and block together (Fig. 10.11b), resolving horizontally gives

$$\sum F_x = 0: \qquad F_p = \mu_1 F_{n1} + F_{n3} \tag{10.22}$$

while resolving vertically gives

$$\sum F_y = 0: \qquad F_{n1} = \mu_3 F_{n3} + F_w \tag{10.23}$$

Eliminating F_{n3} from Eqs. (10.22) and (10.23), we get

$$F_{n1} = \frac{F_w + \mu_3 F_p}{1 + \mu_3 \mu_1} \tag{10.24}$$

Substitution of Eq. (10.24) into (10.21) gives, after rearrangement,

$$F_p = \frac{F_w(\mu_1 + k)}{1 - \mu_3 k} \tag{10.25}$$

where

$$k = \frac{\sin \theta + \mu_2 \cos \theta}{\cos \theta - \mu_2 \sin \theta} \tag{10.26}$$

Now, for example, if θ is 30° and μ_1, μ_2, and μ_3 are equal to 0.2, then

$$k = \frac{0.5 + 0.2(0.866)}{0.866 - 0.2(0.5)} = 0.88$$

and

$$F_p = \frac{F_w(0.2 + 0.88)}{1 - 0.2(0.88)} = 1.3 F_w \tag{10.27}$$

That is, to raise the block, the force applied to the wedge must be greater than 1.3 times the weight of block B.

Repeating the analysis for the case in which the block tends to move downward would give a minimum value of F_p to prevent motion. If this minimum value of F_p were negative, then the block would not move downward when F_p is removed. This would be a "self-locking" system.

10.5   Screw Threads

10.5.1   Square-Threaded Screws

Figure 10.12 shows a square-threaded screw where it can be seen that the nut will be resting on the inclined upper helical surface of the thread. Thus, the thread can be regarded as an inclined plane wrapped around a cylinder, the helix angle α of the thread being equal to the angle of inclination of the plane. The pitch of a screw thread is the distance between corresponding points on successive threads, as shown in Fig. 10.12.

Figure 10.13a shows a section through a square-threaded screw. The pitch p is indicated together with the mean diameter d of the thread. If one revolution of the thread were unwrapped from the cylinder, then an inclined plane would result, as shown in Fig. 10.13b. The nut is represented by the block having the vertical force F_w applied. The horizontal force F_p multiplied by the mean radius $d/2$ would give the moment (or torque) necessary to rotate the nut. The frictional force F_f, parallel to the inclined thread, and the reaction between the nut and thread F_r are also shown.

Resolving forces perpendicular to the inclined plane, we get

$$\sum F_y = 0: \qquad\qquad F_w \cos \alpha + F_p \sin \alpha = F_r \qquad\qquad (10.28)$$

and resolving parallel to the plane, we get

$$\sum F_x = 0: \qquad\qquad F_f + F_w \sin \alpha = F_p \cos \alpha \qquad\qquad (10.29)$$

If the block is sliding or about to slide, then

$$F_f = \mu F_r \qquad\qquad (10.30)$$

where μ is the appropriate coefficient of friction. Eliminating F_f and F_r from Eqs. (10.28)–(10.30), we get

$$\frac{F_p}{F_w} = \frac{\tan \alpha + \mu}{1 - \mu \tan \alpha} \qquad\qquad (10.31)$$

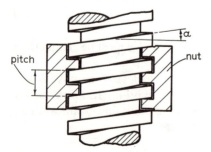

Fig. 10.12   Square-threaded screw.

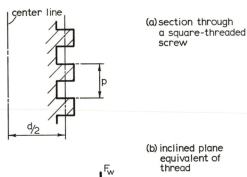

(a) section through a square-threaded screw

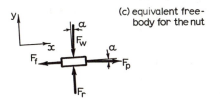

(b) inclined plane equivalent of thread

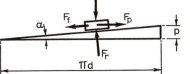

(c) equivalent free-body for the nut

Fig. 10.13 Diagrams for analysis of square-threaded screw.

Now, since $\tan \alpha$ is equal to $p/\pi d$, Eq. (10.31) becomes

$$\frac{F_p}{F_w} = \frac{p + \mu\pi d}{\pi d - \mu p} \tag{10.32}$$

Also, since the moment M required to rotate the nut is given by

$$M = F_p \frac{d}{2} \tag{10.33}$$

then

$$M = \frac{F_w d}{2} \left(\frac{p + \mu\pi d}{\pi d - \mu p}\right) \tag{10.34}$$

Example 10.4 The screw jack shown in Fig. 10.14 has 0.2 thread per millimetre and a mean thread diameter of 100 mm. If the coefficient of friction is 0.3, determine the minimum force F_p to be applied to the bar in order to raise the weight of 5 kN.

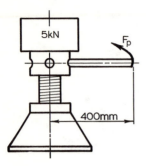

Fig. 10.14 Screw jack.

Solution. The pitch of the thread is given by

$$p = \frac{1}{0.2} = 5 \text{ mm}$$

From Eq. (10.34) the moment required is

$$M = 5000\left(\frac{0.1}{2}\right)\left[\frac{0.005 + 0.3\pi(0.1)}{\pi(0.1) - 0.3(0.005)}\right] = 79.4 \text{ N} \cdot \text{m}$$

Now, since

$$F_p(0.4) = M$$

we get

$$F_p = \frac{M}{0.4} = \frac{79.4}{0.4} = 198 \text{ N}$$

10.5.2   V-Threaded Screws

Figure 10.15 shows a portion of a V-threaded screw where F_r is the normal reaction between the contact surfaces which are supporting an axial force F_w. Resolving vertically, we get

$$\sum F_y = 0: \qquad\qquad F_r \cos \beta = F_w$$

or

$$F_r = \frac{F_w}{\cos \beta}$$

The frictional force F_f (normal to the plane of the paper) is given by

$$F_f = \mu F_r = \frac{\mu F_w}{\cos \beta} \tag{10.35}$$

If the procedure for analyzing a square-threaded screw were now repeated,

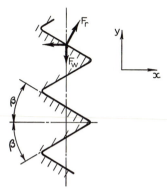

Fig. 10.15 Portion of V-threaded screw.

then wherever μ occurs it would be divided by $\cos \beta$. This means that the equations developed for the square-threaded screw can be used if an equivalent value μ_e for the coefficient of friction is employed, where

$$\mu_e = \frac{\mu}{\cos \beta} \qquad (10.36)$$

Problems

10.1   A rope is coiled around a 0.5-m-diameter ship's capstan. One end of the rope is secured to the dock, and the other is held by a sailor. If the maximum force the sailor can pull on the rope is 250 N, determine the least number of full turns the rope must be wrapped around the capstan if the tension in the rope between the ship and the dock is 15 kN. The coefficient of friction between the rope and the capstan is 0.2.

10.2   A band brake is used to control the speed of the flywheel shown in Fig. P10.2. If the force F_p applied to the lever is 50 N, what torque M_p must be applied to the flywheel in order to maintain constant speed rotation? Assume that the coefficient of dynamic friction between the band and flywheel is 0.3, and neglect the weight of the lever.

10.3   A plate is suspended by a rope passing around the fixed pegs as shown in Fig. P10.3. If the mass of the plate is 30 kg, what force F_p must be applied to the rope to prevent the plate from slipping. The coefficient of friction between the pegs and the rope is 0.40. Assume that the frictional force between each pair of guides and the rope is 10 N.

10.4   A moment M_p of 250 N·m must be applied to wheel A to drive the conveyor belt shown in Fig. P10.4 at a constant velocity. Determine the minimum tensions in the belt for no slipping to occur, and assume that the coefficient of static friction between the conveyor belt and the drive wheel is 0.3.

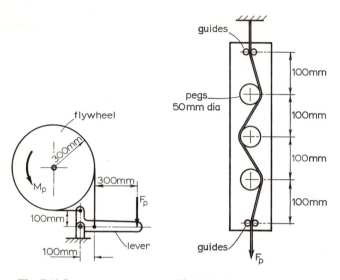

Fig. P10.2   Fig. P10.3

10.5   A 500-N weight is to be raised by using a pulley as shown in Fig. P10.5.
        If the pulley becomes locked, determine (a) the minimum force required to
        hold the load in place and (b) the minimum force required to lift the weight.
        The coefficient of static friction between the rope and the pulley is 0.35.
10.6   Heavy cargo loads are often moved aboard a ship's deck by attaching a rope
        to the load and then wrapping the rope around a motor-driven drum. If
        the rope is wrapped once around the drum, what force must a sailor apply to
        move a 2.5-kN cargo box (Fig. P10.6)? The coefficient of friction between
        the rope and the drum is 0.2, while the coefficient of friction between the
        cargo box and the deck is 0.3.
10.7   A repairman who weighs 600 N is lowered into the large ventilator shaft
        of a ship as shown in Fig. P10.7. The coefficient of friction between the
        rope and the ventilator shaft is 0.5. Neglecting the weight of the rope,
        determine the force F_p required to pull the man out of the ventilator shaft.

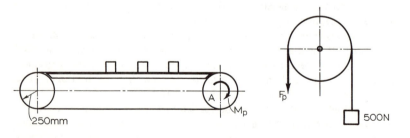

Fig. P10.4   Fig. P10.5

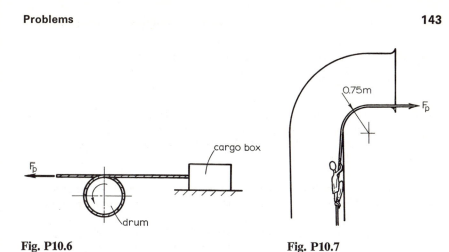

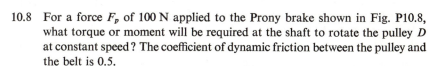

Fig. P10.6   Fig. P10.7

10.8 For a force F_p of 100 N applied to the Prony brake shown in Fig. P10.8, what torque or moment will be required at the shaft to rotate the pulley D at constant speed? The coefficient of dynamic friction between the pulley and the belt is 0.5.

10.9 The conical pivot bearing shown in Fig. P10.9 supports a thrust F_p. Derive an expression for the torque M_p required on the shaft to overcome friction. Assume the pressure between the shaft and the bearing is uniformly distributed.

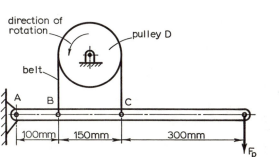

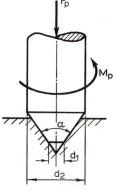

Fig. P10.8   Fig. P10.9

10.10   If a normal load F_n of 250 N is applied to the rotary disk sander shown in
        Fig. P10.10, determine the torque required to rotate the disk. Assume that
        the sander disk is designed so that constant pressure is applied over the area
        of the disk.

10.11   If a metal ball is pressed against a flat plate with a force F_p, the ball flattens
        slightly so that the normal pressure is distributed over a small circle of
        contact having a radius r_0. By the theory of elasticity, the pressure p at a
        distance r from the center of the circle is
        $$p = (3F_p/2\pi r_0{}^2)\sqrt{1 - (r/r_0)^2}$$
        Show that if the ball is rotated slowly about the diametral axis perpendicular
        to the plate while it is pressed against the plate, the required torque or
        moment M_p is given by $(3/16)\pi\mu F_p r_0$, where μ is the coefficient of friction.

10.12   The wedge shown in Fig. P10.12 is used to split wood. If the wedge angle is
        2θ, what must be the value of the coefficient of friction between the wood
        and the wedge for the wedge to be self-locking?

10.13   Two 10° wedges A and B (Fig. P10.13) are positioned so that a horizontal
        force on one wedge can be used to adjust the position of a 5-kN machine
        block C. If the coefficient of friction for all sliding surfaces is 0.2, determine
        the magnitude of the force F_p required to raise the load. Show whether the
        system is self-locking.

10.14   The jack shown in Fig. P10.14 is used to raise and then lower a 5-kN load;

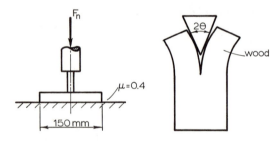

Fig. P10.10   Fig. P10.12

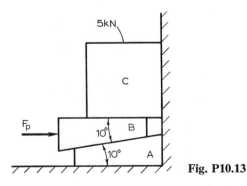

Fig. P10.13

the screws are square-threaded. Determine the torques required to raise and lower the load, respectively, when θ is 45°. The mean diameter of the screw is 20 mm, the pitch is 4 mm, and the coefficient of friction is 0.30. All joints in the frame of the jack are pinned.

10.15 For the screw jack considered in Example 10.4, if the force F_p required to raise the 5-kN weight is removed, will the screw rotate, or is it self-locking?

10.16 The square-threaded screw jack shown in Fig. P10.16 has a pitch of 5 mm and a mean thread diameter of 100 mm. The coefficient of friction for the threads is 0.3. The 5-kN weight is supported on a collar which does not rotate with the screw. If the coefficient of friction between the collar and the top of the jack is 0.1, determine the force F_p required to raise the weight. Assume the top of the jack is lubricated.

10.17 A C-clamp (Fig. P10.17) is used to hold two blocks together. The clamp has a square-threaded screw with a pitch of 3 mm and a mean diameter of 20 mm. Find the torque required if a 2-kN load is required on the blocks and the coefficient of friction for the threads is 0.35. What torque must be applied to loosen the clamp?

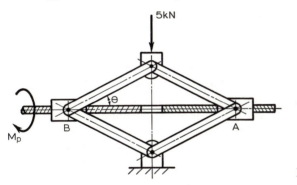

Fig. P10.14

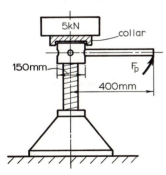

Fig. P10.16   **Fig. P10.17**

11

Hydrostatics

11.1 Introduction

A fluid is sometimes defined as a substance that takes the shape of its container. More precisely, however, a fluid is a continuous substance which when at rest is unable to support shear forces. Thus, a fluid at rest can exert only normal forces on bounding surfaces. The statics of fluids is referred to as hydrostatics when the fluid is a liquid (i.e., a fluid that can be considered incompressible). In this chapter we shall consider the forces on surfaces subject to liquid pressure.

11.2 Pressure Variation in a Liquid

The pressure (force per unit area) at any given point in a liquid at rest is the same in all directions. This is known as Pascal's law and is given by

$$p = p_o + \rho g h \qquad (11.1)$$

where h is the depth of the point from the surface of the liquid, p_o is the pressure at the surface of the liquid (normally atmospheric pressure), and ρ is the density of the liquid (mass per unit volume).

Since atmospheric pressure usually occurs everywhere, Eq. (11.1) is often written as

$$p_g = p - p_o = \rho g h \qquad (11.2)$$

146

where p_g is the difference between the absolute pressure p and the atmospheric pressure p_o. This pressure difference is called *gage pressure*. A negative gage pressure indicates the amount of vacuum.

Example 11.1   A tank filled with a liquid of density ρ is connected to a U-tube manometer filled with mercury of density ρ_m (Fig. 11.1). Determine the pressure at A.

Solution.   The pressures at points B and C are equal since these points are within the same liquid (mercury) and lie at the same level. Now

$$p_B = p_A + \rho g l_1 \qquad (a)$$

$$p_C = p_o + \rho_m g l_2 \qquad (b)$$

Since p_B is equal to p_C, we get

$$p_A = p_o + \rho_m g l_2 - \rho g l_1 \qquad (c)$$

If the density of mercury ρ_m is much greater than that of the liquid in the tank, then

$$p_A \simeq p_o + \rho_m g l_2 \qquad (d)$$

or, the gage pressure is given by

$$p_g = p_A - p_o = \rho_m g l_2 \qquad (e)$$

Example 11.2   Determine the force F_p acting on piston A required to raise the 20-kN load resting on piston B in the hydraulic lift shown in Fig. 11.2.

Solution.   The pressure at point C in Fig. 11.2 is the same in all directions. Thus,

$$p_A = p_C = p_B + \rho g h \qquad (a)$$

If $\rho g h$ is small compared to p_B, then, from Eq. (a),

$$p_A = p_B \qquad (b)$$

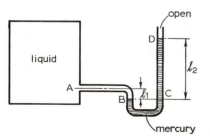

Fig. 11.1   U-tube manometer.

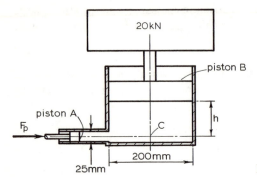

Fig. 11.2 Hydraulic lift.

and since pressure is force divided by area, we get

$$\frac{F_p(4)}{\pi(25)^2} = \frac{20,000(4)}{\pi(200)^2}$$

$$F_p = 312.5 \text{ N}$$

11.3 Hydrostatic Forces on a Submerged Plane Surface

A surface submerged in a liquid, such as a gate valve in a dam, the wall of a tank, or a plate in the submerged hull of a ship, is subjected to liquid pressures normal to the surface. If these pressures are sufficiently large, then it is necessary to determine the magnitude of the resultant force F_r on the surface due to pressure distribution and to locate the point, known as the center of pressure, at which this resultant force may be assumed to act.

Figure 11.3 shows a plane surface submerged in a liquid. The force exerted by the liquid on an elemental strip of area dA is

$$dF_r = p \, dA \tag{11.3}$$

where the pressure p on the elemental strip is given by Eq. (11.1). Thus, substitution of Eq. (11.1) into Eq. (11.3) and integration over the total area give the resultant force F_r:

$$F_r = \int (p_o + \rho g h) \, dA \tag{11.4}$$

Since

$$h = y \sin \theta \tag{11.5}$$

then

$$F_r = \int p_o \, dA + \rho g \sin \theta \int y \, dA \tag{11.6}$$

From Chapter 5 [Eq. (5.11)] the position of the centroid of area y_a is given by

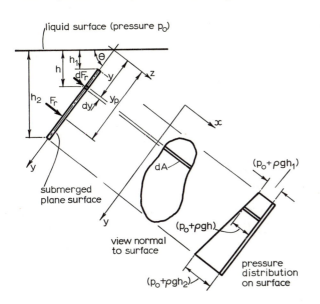

Fig. 11.3 Submerged plane surface.

$$y_a A = \int y \, dA \qquad (11.7)$$

Thus, Eq. (11.6) becomes

$$F_r = p_0 A + \rho g \sin \theta \, y_a A$$

or

$$F_r = p_0 A + \rho g h_a A \qquad (11.8)$$

Hence, the total force on one side of a submerged surface is equal to the total pressure at the centroid of area of the surface multiplied by the total area of the submerged surface. However, since atmospheric pressure acts everywhere, it will produce an equal and opposite force $p_0 A$ on both sides of the submerged surface, its effect cancels, and the effective force, due to liquid pressure, is given by

$$F_r = \rho g h_a A$$
$$= p_{ga} A \qquad (11.9)$$

where p_{ga} is the gage pressure at the centroid of the submerged surface.

The location of the center of pressure (x_p, y_p), that is, the location of the point of application of F_r such that the moment due to the resultant force is the same as the sum of the moments due to the elemental forces, is obtained as follows. First, taking moments about the x axis, we get

$$F_r y_p = \int p_g y \, dA$$

and substitution of Eq. (11.2) and Eq. (11.5) gives

$$F_r y_p = \rho g \sin \theta \int y^2 \, dA \qquad (11.10)$$

By letting

$$I_{xx} = \int y^2 \, dA \qquad (11.11)$$

Eq. (11.10) becomes

$$F_r y_p = \rho g \sin \theta \, I_{xx} \qquad (11.12)$$

or

$$y_p = \frac{\rho g \sin \theta \, I_{xx}}{p_{ga} A} \qquad (11.13)$$

However,

$$p_{ga} = \rho g h_a = \rho g y_a \sin \theta \qquad (11.14)$$

Hence, Eq. (11.13) becomes

$$y_p = \frac{I_{xx}}{y_a A} \qquad (11.15)$$

or, alternatively,

$$y_p = \frac{\int y^2 \, dA}{\int y \, dA} \qquad (11.16)$$

The term I_{xx} is known as the *second moment of area* of the surface about the x axis. Thus, the distance y_p to the center of pressure is equal to the second moment of area of the submerged surface about the x axis divided by the first moment of area about the x axis where the x axis lies in the surface of the liquid and in the plane of the submerged surface.

A similar procedure can be used to determine the x coordinate of the center of pressure. Again, considering the effect of the gage pressure and taking moments about the y axis, we get

$$F_r x_p = \int \rho g h x \, dA = \rho g \sin \theta \int xy \, dA \qquad (11.17)$$

By letting

$$I_{xy} = \int xy \, dA \qquad (11.18)$$

Eq. (11.17) becomes

$$x_p = \frac{I_{xy}}{y_a A} \qquad (11.19)$$

or, alternatively,

$$x_p = \frac{\int xy \, dA}{\int y \, dA} \qquad (11.20)$$

The term I_{xy} is called the *product of inertia* of the area with respect to the x, y axes, and the distance x_p to the center of pressure is equal to the product of inertia of the area with respect to the x, y axes divided by the first moment of area about the x axis where the x and y axes lie in the plane of the submerged surface.

Example 11.3   Determine the resultant force exerted on the semicircular end of the water tank shown in Fig. 11.4a if the tank is filled to capacity. Express the result in terms of the radius r of the semicircular end and the density ρ of the water.

Solution.   The resultant force due to liquid pressure on the end of the tank is given by

$$F_r = \int p_g \, dA = \int \rho g y \, dA \tag{a}$$

where y is the depth below the surface of the water. The elemental area dA is given by (Fig. 11.3b)

$$dA = 2r \cos \theta \, dy \tag{b}$$

Since

$$y = r \sin \theta \tag{c}$$

then

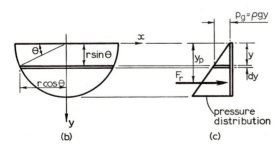

(a)

(b)

(c)

Fig. 11.4   Diagrams for Example 11.3.

$$dy = r \cos \theta \, d\theta \qquad (d)$$

and thus combining Eqs. (a)–(d) gives

$$F_r = 2\rho g r^3 \int_0^{\pi/2} \sin \theta \cos^2 \theta \, d\theta$$
$$= 2\rho g r^3 (-\tfrac{1}{3} \cos^3 \theta)_0^{\pi/2}$$
$$= \tfrac{2}{3} \rho g r^3 \qquad (e)$$

The resultant force F_r acts at the center of pressure, the location of which is determined by equating the moment due to F_r to the sum of the moments due to the elemental forces. Taking moments about the x axis gives

$$F_r y_p = \int \rho g y^2 \, dA$$
$$= 2\rho g r^4 \int_0^{\pi/2} \sin^2 \theta \cos^2 \theta \, d\theta$$
$$= 2\rho g r^4 \int_0^{\pi/2} (\sin^2 \theta - \sin^4 \theta) \, d\theta$$
$$= \frac{\pi \rho g r^4}{8} \qquad (f)$$

Substitution of Eq. (e) into Eq. (f) gives

$$y_p = \frac{3\pi r}{16} \qquad (g)$$

Alternatively, these same results can be obtained directly from Eqs. (11.9) and (11.15). In this case,

$$A = \frac{\pi r^2}{2}$$

$$y_a = \frac{4}{3} \frac{r}{\pi}$$

Thus, from Eq. (11.9),

$$F_r = p_{ga} A = \rho g y_a A$$

or

$$F_r = \rho g \frac{4}{3} \frac{r}{\pi} \left(\frac{\pi}{2} r^2 \right)$$

$$= \tfrac{2}{3} \rho g r^3$$

Since

$$I_{xx} = \int y^2 \, dA$$
$$= 2 r^4 \int_0^{\pi/2} \sin^2 \theta \cos^2 \theta \, d\theta$$
$$= \frac{\pi r^4}{8}$$

then, from Eq. (11.15),

$$y_p = \frac{I_{xx}}{y_a A}$$

$$= \frac{\pi r^4/8}{\frac{4}{3}(r/\pi)(\pi r^2/2)} = \frac{3\pi r}{16}$$

11.4 Second Moment of Area

The concept of second moment of area (sometimes known as the area moment of inertia) was introduced in the previous section. This property of an area is a common parameter in engineering problems.

The second moment of area I is always related to a particular axis and is the sum of the second moments of the elements of the area about the specified axis. Hence, for a plane area as shown in Fig. 11.5, the second moment of area about the x-x axis is given by

$$I_{xx} = \int y^2 \, dA \tag{11.21}$$

If the second moment of area about an axis a-a passing through the centroid of the area is denoted by I_{aa} (Fig. 11.6), then the second moment of area about an axis x-x parallel to a-a and a distance d from a-a is

$$I_{xx} = \int (d - y)^2 \, dA$$
$$= \int (d^2 - 2dy + y^2) \, dA$$
$$= \int d^2 \, dA - 2d \int y \, dA + \int y^2 \, dA$$
$$= Ad^2 - 2d \int y \, dA + I_{aa}$$

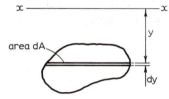

Fig. 11.5   Plane area.

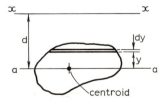

Fig. 11.6   Parallel axes.

or, since $\int y \, dA$ is zero when y is measured from the centroid of the area, we get

$$I_{xx} = I_{aa} + Ad^2 \tag{11.22}$$

Thus, if I_{aa} is known, the second moment of area about any axis x-x parallel to a-a and distance d from it can be found from Eq. (11.22); this is known as the parallel-axis theorem.

Example 11.4 Determine the second moments of area of the rectangle shown in Fig. 11.7a about the a-a and x-x axes.

Solution. Figure 11.7b shows an elemental strip of area $b \, dy$ a distance y from the a-a axis. Thus,

$$I_{aa} = \int y^2 \, dA$$
$$= \int_{-l/2}^{l/2} y^2 b \, dy = b \left(\frac{y^3}{3} \right)_{-l/2}^{l/2}$$
$$= \frac{bl^3}{12}$$

Now, using Eq. (11.22), we get

$$I_{xx} = I_{aa} + Ad^2$$
$$= \frac{bl^3}{12} + bl \left(\frac{l}{2} \right)^2$$
$$= \frac{bl^3}{3}$$

Example 11.5 Determine the second moment of area for the circular area shown in Fig. 11.8a about a centroidal axis (axis passing through the centroid of area) lying in the plane of the paper.

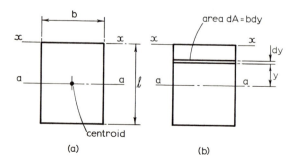

(a)

(b)

Fig. 11.7 Figure for Example 11.4.

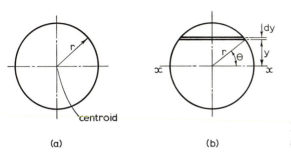

Fig. 11.8 Figure for Example 11.5.

(a)   (b)

Solution. The elemental strip shown in Fig. 11.8b has an area of $2x\,dy$, so that

$$I_{xx} = 2 \int y^2 x \, dy \qquad (a)$$

Since

$$x = r \cos \theta$$
$$y = r \sin \theta \qquad (b)$$

and

$$dy = r \cos \theta \, d\theta \qquad (c)$$

Eq. (a) becomes

$$I_{xx} = 2 \int_{-\pi/2}^{\pi/2} r^4 \sin^2 \theta \cos^2 \theta \, d\theta$$
$$= 2r^4 \int_{-\pi/2}^{\pi/2} \sin^2 \theta (1 - \sin^2 \theta) \, d\theta$$
$$= \frac{\pi r^4}{4} \qquad (d)$$

Example 11.6 Determine the second moment of area for the area shown in Fig. 11.9a about the centroidal axis which is parallel to the 3-m side.

Solution. To determine the location of the centroidal axis *a-a*, the area is divided into two rectangular sections (1) and (2) (Fig. 11.9b). From Chapter 5, the distance of the centroidal axis from the axis *x-x* is given by

$$y_a = \frac{A_1 y_{a1} + A_2 y_{a2}}{A_1 + A_2}$$

$$= \frac{3(1)(2.5) + 2(1)(1)}{3(1) + 2(1)} = 1.90 \text{ m}$$

The second moment of area of section (1) about the *a-a* axis is given by

$$I_{aa1} = \frac{b_1 l_1^3}{12} + A_1 d_1^2 = \frac{3(1)^3}{12} + 3(1)(0.6)^2$$

$$= 1.33 \text{ m}^4$$

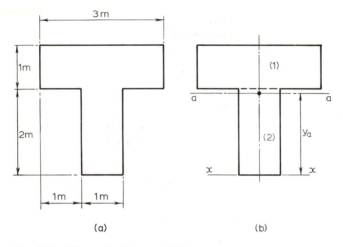

Fig. 11.9 Figure for Example 11.6.

The second moment of area of section (2) about the a-a axis is

$$I_{aa2} = \frac{b_2 l_2^3}{12} + A_2 d_2^2 = \frac{1(2)^3}{12} + 2(1)(0.9)^2$$

$$= 2.29 \text{ m}^4$$

Thus,

$$I_{aa} = I_{aa1} + I_{aa2}$$

$$= 1.33 + 2.29$$

$$= 3.62 \text{ m}^4$$

11.5 Product of Inertia of an Area

The product of inertia of an area was introduced earlier and with respect to the x, y axes is defined by the expression

$$I_{xy} = \int xy \, dA \tag{11.23}$$

Unlike the second moment of area which is always positive, the product of inertia of an area may be positive or negative. The product of inertia of an area is zero whenever one of the reference axes is an axis of symmetry. For example, for the area shown in Fig. 11.10 the y axis is an axis of symmetry, and for each element of area located at $+x$, $+y$ there is an equal element of

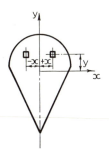

Fig. 11.10 Example where the product of inertia of area is zero with respect to the x, y axes.

area located at $-x$, $+y$, and for each pair of elements the products xy cancel each other. Thus, the product of inertia for the whole area is zero.

A parallel-axis theorem, similar to that for the second moment of area, also applies to the product of inertia of an area. To derive this theorem we can consider the area shown in Fig. 11.11 where the x', y' axis system is a centroidal axis system parallel to the x, y axis system. The product of inertia of the area, with respect to the x, y axes, is

$$I_{xy} = \int (x' + x_a)(y' + y_a)\, dA$$
$$= \int x'y'\, dA + x_a \int y'\, dA + y_a \int x'\, dA + x_a y_a \int dA$$

Now, since the second and third terms are both zero, we get

$$I_{xy} = I_{x'y'} + x_a y_a A \qquad (11.24)$$

where $I_{x'y'}$ is the product of inertia of the area with respect to the x', y' axes.

Example 11.7 For the rectangular area shown in Fig. 11.12, determine the products of inertia of the area with respect to x', y' centroidal axis system and with respect to the x, y axis system.

Solution. Because the x', y' axes are axes of symmetry,

$$I_{x'y'} = 0$$

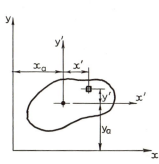

Fig. 11.11 Parallel axes.

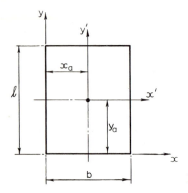

Fig. 11.12 Diagram for Example 11.7.

To determine I_{xy}, use is made of Eq. (11.24). Thus,

$$I_{xy} = I_{x'y'} + x_a y_a A$$

$$= 0 + \left(\frac{b}{2}\right)\left(\frac{l}{2}\right)bl$$

$$= \frac{b^2 l^2}{4}$$

Problems

11.1 The differential manometer shown in Fig. P11.1 indicates the difference in the pressures within tanks (1) and (2). Obtain an expression for this pressure difference in terms of the weight densities γ_1, γ_2, and γ_m of the liquids in the tanks and manometer, respectively, and the distances l_1, l_2, and l_3.

11.2 A barometer is a device used for measuring atmospheric pressure (Fig. P11.2). It can be constructed by filling the closed-end tube with air-free mercury, inverting the tube, and inserting it in the reservoir with its open end beneath the surface of the mercury. Neglecting the small pressure of mercury vapor above the column, determine the height h of the mercury

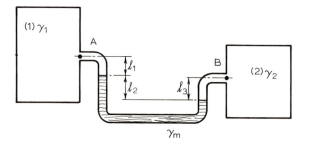

Fig. P11.1

      column when the atmospheric pressure is 101.33 kN/m². The mass density of mercury is 13,555 kg/m³ at 15.6 °C.

11.3   Determine the force F_p required to raise the 25-kN load resting on piston B shown in Fig. P11.3.

11.4   An Olympic-sized swimming pool is shown in Fig. P11.4. Determine the resultant forces and the locations of these forces on the bottom of the pool, on the ends of the pool, and on the window in the side of the pool. The mass density of water is 1000 kg/m³.

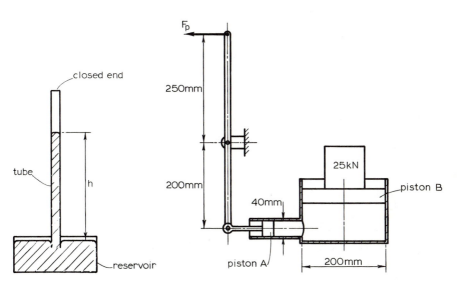

Fig. P11.2   **Fig. P11.3**

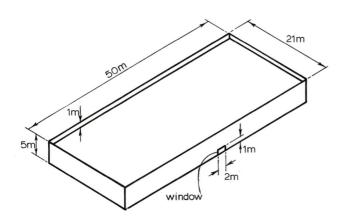

Fig. P11.4

11.5 The vertical door shown in Fig. P11.5 is used to close the end of a rectangular water channel 3 m wide. If the door is hinged at A, at what depth of water h will the door open? The mass density of water is 1000 kg/m³.

11.6 Figure P11.6 shows an automatic value which consists of a square plate 0.27 m × 0.27 m which is pivoted about a horizontal axis through A located at a distance 0.12 m above its lower edge. Determine the depth of water h for which the valve will open. The mass density of water is 1000 kg/m³.

11.7 Determine the second moment of area of the triangular section shown in Fig. P11.7 about the axis x-x.

11.8 Determine the second moment of area for the I-section shown in Fig. P11.8 about a horizontal axis passing through its centroid and about a vertical axis passing through its centroid.

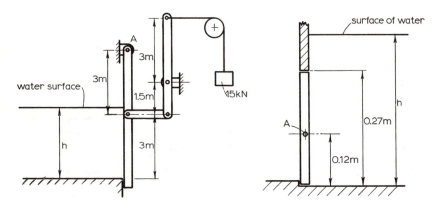

Fig. P11.5   **Fig. P11.6**

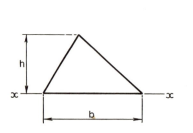

Fig. P11.7

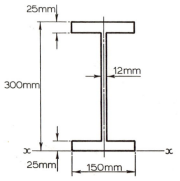

Fig. P11.8

11.9   Determine the second moment of area for the I-section shown in Fig. P11.8 about the *x-x* axis.

11.10   Determine the product of inertia I_{xy} of the area shown in Fig. P11.10.

11.11   Determine the second moment of area for the channel section shown in Fig. P11.11 about the horizontal and vertical centroidal axes.

11.12   For the section shown in Fig. P11.12, determine the magnitude of *d* such that I_{xx} is equal to I_{yy}.

11.13   Find the position of the centroid and the second moment of area about the axis *a-a* for the area shown in Fig. P11.13.

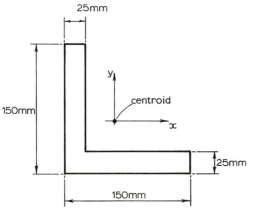

Fig. P11.10

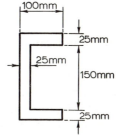

Fig. P11.11

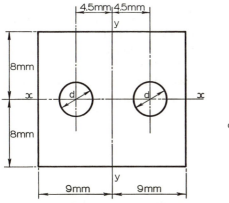

Fig. P11.12

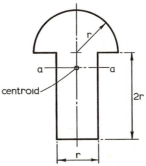

Fig. P11.13

12

Kinematics of Particles

12.1 Introduction

The general study of the relationships among displacement, length, and time is called kinematics. Basically, it is a study of the motion of particles and rigid bodies without regard to the forces causing the motion; it deals with displacement, velocity, acceleration, and time and is often referred to as the geometry of motion.

12.2 Rectilinear Motion

Rectilinear motion of a particle (or point) is defined as the motion of the particle along a straight path. Figure 12.1 shows such a particle, represented by the point P, which is moving along a straight path denoted by the line Os. At a certain time t_1 the point is located a distance s_1 from the stationary reference point O, and at a later time t_2 the point is located a distance s_2 from the point O. The time interval under consideration is therefore $t_2 - t_1$ and can be

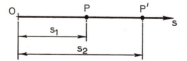

Fig. 12.1 Linear displacement of a particle P.

162

represented by the symbol Δt. Similarly, the displacement of the point is $s_2 - s_1$ or Δs.

The instantaneous velocity v of the point is defined as the rate of change of its position; its average velocity v_{av} during the time interval under consideration can be obtained by dividing the displacement by the time interval. Thus,

$$v_{av} = \frac{\Delta s}{\Delta t} \tag{12.1}$$

If the average velocity is measured over an infinitesimally short time interval (i.e., at a given instant) so that Δt approaches zero ($\Delta t \rightarrow 0$), then it is known as the instantaneous velocity v. Thus at time t, v is given by $\Delta s/\Delta t$ in the limit as Δt approaches zero, which is designated ds/dt. Hence,

$$v = \lim_{\Delta t \rightarrow 0} \frac{\Delta s}{\Delta t} = \frac{d(s)}{dt} = \frac{ds}{dt} = \dot{s} \tag{12.2}$$

Figure 12.2a, where distance s is plotted as a function of time t, shows this graphically, and it can be seen that as Δt approaches zero the instantaneous velocity will be the slope of the graph. The instantaneous acceleration a of the point is defined as the rate of change of its velocity. Thus,

$$a = \lim_{\Delta t \rightarrow 0} \frac{\Delta v}{\Delta t} = \frac{d(v)}{dt} = \frac{dv}{dt} = \dot{v} \tag{12.3}$$

Now, since v is equal to ds/dt,

$$a = \frac{d}{dt}\left(\frac{ds}{dt}\right) = \frac{d^2s}{dt^2} = \ddot{s} \tag{12.4}$$

Figure 12.2b illustrates that the acceleration a is equal to the slope of the graph of velocity v plotted as a function of time t when Δt approaches zero. If Eq. (12.2) is rearranged and then integrated, we get

$$\int_{t_1}^{t_2} v \, dt = \int_{s_1}^{s_2} ds = s_2 - s_1 = \Delta s \tag{12.5}$$

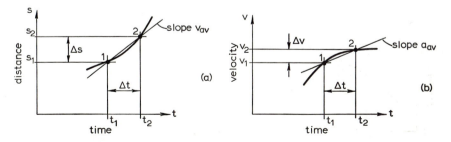

Fig. 12.2   Relations among displacement, velocity, and time in rectilinear motion.

Thus, the displacement Δs of the point is given by the area under the velocity-time curve (Fig. 12.3a).

Similarly, from Eq. (12.3),

$$\int_{t_1}^{t_2} a \, dt = \int_{v_1}^{v_2} dv = v_2 - v_1 = \Delta v \qquad (12.6)$$

and the change in velocity Δv of the point is given by the area under the acceleration-time curve (Fig. 12.3b).

When the acceleration of a point is known as a function of the distance s rather than time t, it may be more convenient to rewrite Eq. (12.3) as follows:

$$a = \frac{dv}{ds}\frac{ds}{dt} = v\frac{dv}{ds} \qquad (12.7)$$

In this case, if the acceleration a is plotted as a function of the distance s, the area under the curve during the interval from s_1 to s_2 is (Fig. 12.4)

$$\int_{s_1}^{s_2} a \, ds = \int_{v_1}^{v_2} v \, dv = \tfrac{1}{2}(v_2^2 - v_1^2) \qquad (12.8)$$

Example 12.1   An automobile starts from rest with an initial acceleration of 5 m/s^2. If the acceleration falls uniformly with increased velocity until it

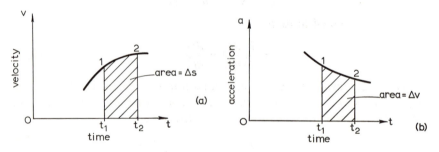

Fig. 12.3   Relations among velocity, acceleration, and time.

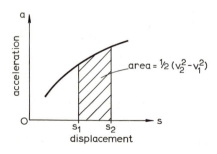

Fig. 12.4   Relation between acceleration and distance.

becomes zero when the velocity of the car reaches 50 m/s, how long does it take for the car to reach a velocity of 30 m/s?

Solution. Figure 12.5 shows the relation between acceleration and velocity. This relation can be expressed by

$$a = 5 - 0.1v \qquad \text{(a)}$$

Now

$$a = \frac{dv}{dt} = 5 - 0.1v \qquad \text{(b)}$$

After rearrangement and integration, we get

$$\int_{t_1}^{t_2} dt = \int_{v_1}^{v_2} \frac{dv}{5 - 0.1v}$$

$$t_2 - t_1 = -\left[\frac{\ln(5 - 0.1v)}{0.1}\right]_{v_1}^{v_2}$$

$$= -10[\ln(5 - 0.1v_2) - \ln(5 - 0.1v_1)]$$

$$= -10\ln\left(\frac{5 - 0.1v_2}{5 - 0.1v_1}\right) \qquad \text{(c)}$$

Now, when $t_1 = 0$, $v_1 = 0$, and $t = t_2$, $v_2 = 30$ m/s. Thus,

$$\Delta t = -10\ln\left[\frac{5 - 0.1(30)}{5}\right] = 9.16 \text{ s}$$

12.2.1 Special Case of Constant Acceleration

When a particle falls under the action of gravity and where the air resistance is negligible, the particle is subject to a constant acceleration. This is one example of motion at constant acceleration. When the acceleration a is constant, Eq. (12.6) becomes

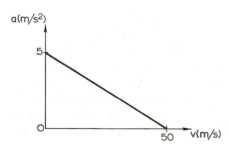

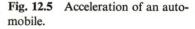

Fig. 12.5 Acceleration of an automobile.

$$a \int_{t_1}^{t_2} dt = v_2 - v_1$$

or
$$v_2 = v_1 + a(t_2 - t_1) \tag{12.9}$$

Similarly, Eq. (12.8) becomes

$$a \int_{s_1}^{s_2} ds = \tfrac{1}{2}(v_2^2 - v_1^2)$$

or
$$v_2^2 = v_1^2 + 2a(s_2 - s_1) \tag{12.10}$$

Finally, eliminating v_1 from Eqs. (12.9) and (12.10), we get

$$s_2 - s_1 = v_1(t_2 - t_1) + \frac{a}{2}(t_2 - t_1)^2 \tag{12.11}$$

Example 12.2   When the engine in a model rocket ceases to burn, the rocket is traveling vertically upward at a velocity of 70 m/s and has an altitude of 100 m. What is the maximum altitude gained by the rocket if air resistance is neglected?

Solution.   Using Eq. (12.10) and writing $s_1 = 100$ m, $v_1 = 70$ m/s, $v_2 = 0$, and $a = -g = -9.81$ m/s², we get

$$v_2^2 = v_1^2 + 2a(s_2 - s_1)$$
$$0 = (70)^2 - 2(9.81)(s_2 - 100)$$

or the maximum altitude is given by

$$s_2 = \frac{(70)^2}{2(9.81)} + 100$$
$$= 350 \text{ m}$$

Example 12.3   An aircraft lands on the deck of an aircraft carrier and is brought to rest in 3 s by the arrestor gear of the carrier. An accelerometer attached to the aircraft provides a record of the deceleration as shown in Fig. 12.6. Determine (a) the initial velocity of the aircraft relative to the carrier and (b) the distance the aircraft travels along the deck of the aircraft carrier. Assume the carrier is motionless.

Solution.   The best approach with this type of problem is to tabulated the calculations, as shown in Table 12.1. Since it is known that the velocity v is zero after 3 s, we can work backwards from a time t of 3 s until we have found the initial velocity. Referring to Table 12.1, column (1) gives the time t and column (2) gives the acceleration a for each value of t. Column (3) lists the average acceleration a_{av} for each time interval, and since the change in velocity Δv for each interval is given by $a_{av} \Delta t$, we can now find Δv; these values are listed in column (4). Now, working backwards from a final velocity of zero

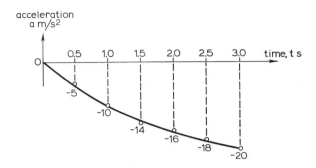

Fig. 12.6 Deceleration of aircraft.

Table 12.1 Tabular Solution to Example 12.3.

| (1) | (2) | (3) | (4) | (5) | (6) | (7) |
| | | | $\Delta v =$ | | | $\Delta s =$ |
t	$-a$	$-a_{av}$	0.5(3)	v	v_{av}	0.5(6)
3.0	20			0		
		19	9.5		4.75	2.38
2.5	18			9.5		
		17	8.5		13.75	6.88
2.0	16			18.0		
		15	7.5		21.75	10.88
1.5	14			25.5		
		12	6.0		28.50	14.25
1.0	10			31.5		
		7.5	3.75		33.38	16.69
0.5	5			35.3		
		2.5	1.25		35.38	17.94
0	0			36.5		
				Initial velocity (m/s)		$\Sigma \, \Delta s =$ 69.02 m

and adding successive values of Δv, we can find the initial velocity; this is done in column (5). Column (6) gives the average velocity v_{av} for each time interval, and column (7) gives the distance Δs traveled during each time interval, since Δs is given by $v_{av} \, \Delta t$. Finally, summing the values of Δs in column (7) gives the total distance traveled.

12.3 Rotational Motion

Figure 12.7 shows a plane area which rotates about the point O. At time t_1 the line OA drawn on the plane area makes an angle θ_1 with a fixed reference line. At time t_2 the line moves to OA', which lies at an angle θ_2 to the reference line. As before, $t_2 - t_1$ is the time interval Δt, and $\theta_2 - \theta_1$ is the change in angular position $\Delta\theta$.

The instantaneous angular velocity ω, which is the rate of change of angular position, is given by

$$\omega = \lim_{\Delta t \to 0} \frac{\Delta\theta}{\Delta t} = \frac{d\theta}{dt} = \dot{\theta} \tag{12.12}$$

Similarly, the angular acceleration α, which is the rate of change of angular velocity, is given by

$$\alpha = \lim_{\Delta t \to 0} \frac{\Delta\omega}{\Delta t} = \frac{d\omega}{dt} = \frac{d^2\theta}{dt^2} = \ddot{\theta} \tag{12.13}$$

Equations (12.12) and (12.13) are identical to Eqs. (12.2) and (12.4) except that the angular velocity ω has been substituted for the linear velocity v, the angular acceleration α has been substituted for the linear acceleration a, and the angular position θ has been substituted for the linear position s. Thus, the graphical relationships and mathematical procedures for the variables s, v, a, and t can also be used for the variables θ, ω, α, and t, respectively, by making the appropriate substitutions.

Example 12.4 Figure 12.8 shows an airplane traveling overhead at a constant altitude h and at a constant horizontal velocity v. Determine the angular velocity ω and angular acceleration α of the line-of-sight tracking device on the ground. The angular position is measured from the vertical at the tracking device.

Solution. From Fig. 12.9, which shows the geometry, we get

$$x = h \tan \theta \tag{a}$$

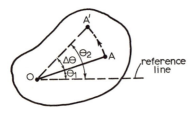

Fig. 12.7 Rotation of a plane area.

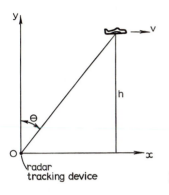

Fig. 12.8 Airplane traveling at constant altitude.

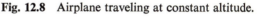

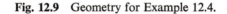

Fig. 12.9 Geometry for Example 12.4.

Differentiation of Eq. (a) gives

$$v = \dot{x} = h \sec^2 \theta \, \dot{\theta}$$

or

$$\dot{\theta} = \omega = \left(\frac{v}{h}\right) \cos^2 \theta \tag{b}$$

Differentiation of Eq. (b) gives

$$\ddot{\theta} = -2\left(\frac{v}{h}\right) \cos \theta \sin \theta \, \dot{\theta} \tag{c}$$

and substitution of $\dot{\theta}$ from Eq. (b) into Eq. (c) gives

$$\ddot{\theta} = \alpha = -\left(\frac{v}{h}\right)^2 \sin 2\theta \cos^2 \theta \tag{d}$$

Example 12.5 Figure 12.10 shows a Scotch yoke mechanism which is used to convert rotary motion to reciprocating motion. As the crank rotates, the pin slides in the slot and pulls the yoke backward and forward. Obtain expressions for the linear velocity and acceleration of the yoke in terms of the angular displacement θ and the constant angular velocity ω of the crank.

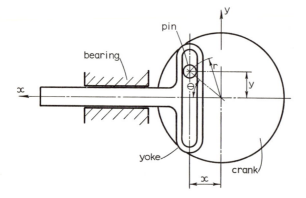

Fig. 12.10   Scotch yoke.

Solution.   From Fig. 12.10, by geometry,

$$x = r \cos \theta \qquad \qquad \text{(a)}$$

Differentiation of Eq. (a) twice with respect to t gives

$$v = \dot{x} = -r \sin \theta \, \dot{\theta} = -\omega r \sin \theta \qquad \qquad \text{(b)}$$

$$a = \ddot{x} = -r \cos \theta \, \dot{\theta}^2 = -\omega^2 r \cos \theta \qquad \qquad \text{(c)}$$

Equation (b) shows that, at the end of the stroke when $\theta = 0$, $\dot{x} = 0$. Equation (c) shows that, although the velocity is zero at the end of the stroke, the acceleration is a maximum.

Substitution of Eq. (a) into Eq. (c) gives

$$a = -\omega^2 x \qquad \qquad \text{(d)}$$

which means that, since ω^2 is a constant, the acceleration of the yoke is proportional to its displacement from its mean position and in the opposite sense. This type of motion is known as *simple harmonic motion* and will be dealt with in more detail later.

12.4   Plane Curvilinear Motion

The motion of a particle (or point) along a curved path which is restricted to a single plane is called *plane curvilinear motion*. In this case, two dimensions are required to define the position of the point, and therefore this type of motion has two degrees of freedom, one more than the case of rectilinear motion.

12.4.1 Rectangular Coordinates

Figure 12.11 shows point P which is moving along a curved path in the xy plane. At time t_1, P is located at the coordinates (x_1, y_1), while at time t_2 it has coordinates (x_2, y_2). Thus, the components of velocity parallel to the x and y axes are, respectively,

$$v_x = \lim_{\Delta t \to 0} \frac{x_2 - x_1}{t_2 - t_1} = \lim_{\Delta t \to 0} \frac{\Delta x}{\Delta t} = \frac{dx}{dt} = \dot{x} \tag{12.14}$$

$$v_y = \lim_{\Delta t \to 0} \frac{y_2 - y_1}{t_2 - t_1} = \lim_{\Delta t \to 0} \frac{\Delta y}{\Delta t} = \frac{dy}{dt} = \dot{y} \tag{12.15}$$

Since the velocity of P can also change during this time interval, the components of acceleration are

$$a_x = \lim_{\Delta t \to 0} \frac{\Delta v_x}{\Delta t} = \frac{dv_x}{dt} = \ddot{x} \tag{12.16}$$

$$a_y = \lim_{\Delta t \to 0} \frac{\Delta v_y}{\Delta t} = \frac{dv_y}{dt} = \ddot{y} \tag{12.17}$$

12.4.2 Polar Coordinates

When the point P is moving in a curved path, it is sometimes more convenient to express the curvilinear motion of P in terms of (1) a distance r along a radial line from the origin of the rectangular coordinate system to the instantaneous position of P and (2) the angle θ that the radial line makes with the x axis (Fig. 12.12). This coordinate system is called a polar coordinate system. Hence, in terms of the polar coordinates r and θ, the position of P is given by

$$x = r \cos \theta \tag{12.18}$$

$$y = r \sin \theta \tag{12.19}$$

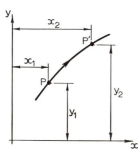

Fig. 12.11 Curvilinear motion in rectangular coordinates.

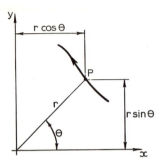

Fig. 12.12   Curvilinear motion in polar coordinates.

Differentiation of Eqs. (12.18) and (12.19) twice with respect to time gives

$$v_x = \dot{x} = -r \sin \theta \, \dot{\theta} + \dot{r} \cos \theta \tag{12.20}$$

$$v_y = \dot{y} = r \cos \theta \, \dot{\theta} + \dot{r} \sin \theta \tag{12.21}$$

and

$$a_x = \ddot{x} = -r\ddot{\theta} \sin \theta - r\dot{\theta}^2 \cos \theta - 2\dot{r}\dot{\theta} \sin \theta + \ddot{r} \cos \theta \tag{12.22}$$

$$a_y = \ddot{y} = r\ddot{\theta} \cos \theta - r\dot{\theta}^2 \sin \theta + 2\dot{r}\dot{\theta} \cos \theta + \ddot{r} \sin \theta \tag{12.23}$$

By projecting the rectangular components of velocity (v_x and v_y) and acceleration (a_x and a_y) onto the r and θ directions (Fig. 12.13) we get the radial and angular components of velocity v_r and v_θ, respectively. Thus,

$$v_r = v_x \cos \theta + v_y \sin \theta \tag{12.24}$$

$$v_\theta = v_y \cos \theta - v_x \sin \theta \tag{12.25}$$

Substitution of Eqs. (12.20) and (12.21) into Eqs. (12.24) and (12.25) gives

$$v_r = \dot{r}(\cos^2 \theta + \sin^2 \theta) - r\dot{\theta}(\sin \theta \cos \theta - \cos \theta \sin \theta) = \dot{r} \tag{12.26}$$

$$v_\theta = \dot{r}(\cos \theta \sin \theta - \sin \theta \cos \theta) + r\dot{\theta}(\sin^2 \theta + \cos^2 \theta) = r\dot{\theta} \tag{12.27}$$

Thus, the velocity of P can be expressed in terms of the rate of change of the

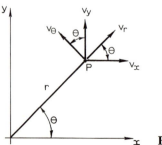

Fig. 12.13   Components of velocity.

length r of the line OP and the rate of change of angular position θ of this same line.

In a similar manner we can get the radial and angular components of acceleration a_r and a_θ, respectively. Thus,

$$a_r = a_x \cos \theta + a_y \sin \theta = \ddot{r} - r\dot{\theta}^2 \tag{12.28}$$

$$a_\theta = a_y \cos \theta - a_x \sin \theta = r\ddot{\theta} + 2\dot{r}\dot{\theta} \tag{12.29}$$

12.4.3 Tangential Coordinates

The velocity and acceleration of a point moving along a curved path can also be expressed in terms of components tangent to the curved path and normal to the path. Figure 12.14 shows the situation being considered. The position of P is specified by its distance s along the curved path and measured from a given reference point on the path. Therefore, in this coordinate system, the velocity components tangent to and normal to the path, v_t and v_n, respectively, are

$$v_t = \dot{s}$$

$$v_n = 0 \tag{12.30}$$

The corresponding components of acceleration are determined by considering the change in the components of velocity tangent and normal to the path as the point moves from P to P'. If the tangential velocity at time t_1 is v_t and the tangential velocity at time t_2 is $v_t + \Delta v$, then a_t, the component of acceleration tangent to the path, is

$$a_t = \lim_{\Delta t \to 0} \frac{(v_t + \Delta v) \cos \Delta \theta - v_t}{\Delta t}$$

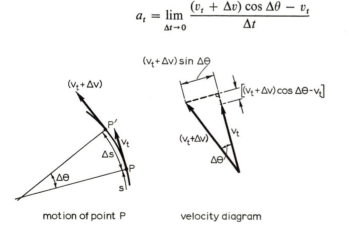

motion of point P   velocity diagram

Fig. 12.14 Curvilinear motion in tangential coordinates.

Since $\Delta\theta$ is small, $\cos \Delta\theta \approx 1$, and we get

$$a_t = \lim_{\Delta t \to 0} \frac{\Delta v_t}{\Delta t} = \frac{dv_t}{dt} = \ddot{s} \tag{12.31}$$

Also, a_n, the component of acceleration normal to the path, is

$$a_n = \lim_{\Delta t \to 0} \frac{(v_t + \Delta v) \sin \Delta\theta}{\Delta t}$$

Since $\Delta\theta$ is small, $\sin \Delta\theta \approx \Delta\theta$, and we get

$$a_n = \lim_{\Delta t \to 0} \left(v_t \frac{\Delta\theta}{\Delta t} + \frac{\Delta v \, \Delta\theta}{\Delta t} \right)$$

Finally, since $\Delta v \, \Delta\theta$ is a higher-order term, it can be neglected, and thus

$$a_n = v_t \dot{\theta} = \dot{s} \dot{\theta} \tag{12.32}$$

By introducing the instantaneous radius of curvature ρ of the path and noting that ds is equal to $\rho \, d\theta$ (or $\dot{s} = \rho\dot{\theta}$) then a_n can be expressed in the more useful form

$$a_n = \frac{\dot{s}^2}{\rho} \tag{12.33}$$

Example 12.6 The motion of the collar B mounted on the rotating rod OA (Fig. 12.15a) is defined by the equations

$$r = 2t^2 - t^3$$
$$\theta = 3t^2$$

where r is measured in metres, θ in radians, and t in seconds. Determine the radial and angular components of the velocity and acceleration of the collar when t is 1 s.

Solution. Differentiating the equations for r and θ twice gives

$$\dot{r} = 4t - 3t^2$$
$$\ddot{r} = 4 - 6t$$
$$\dot{\theta} = 6t$$
$$\ddot{\theta} = 6$$

Referring to Fig. 12.15b and using Eqs. (12.26) and (12.27), we get

$$v_r = \dot{r} = 4t - 3t^2 = 1 \text{ m/s}$$
and
$$v_\theta = r\dot{\theta} = (2t^2 - t^3)(6t) = 6 \text{ m/s}$$

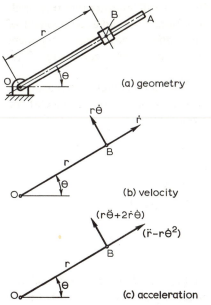

(a) geometry

(b) velocity

(c) acceleration   **Fig. 12.15** Diagrams for Example 12.6.

Referring to Fig. 12.15c and using Eqs. (12.28) and (12.29), we get

$$a_r = \ddot{r} - r\dot{\theta}^2 = (4 - 6t) - (2t^2 - t^3)(6t)^2 = -38 \text{ m/s}^2$$
$$a_\theta = r\ddot{\theta} + 2\dot{r}\dot{\theta} = (2t^2 - t^3)(6) + 2(4t - 3t^2)(6t) = 18 \text{ m/s}^2$$

12.5 Motion Along a Circular Path

In many practical situations, the curve along which a point moves is a circular arc. For example, any point in a rigid body which is rotating about a fixed axis is moving in a circular path. For a point moving in a circular path, the radius r from the axis of rotation is constant, and therefore $\dot{r}$ and $\ddot{r}$ are zero, and for polar coordinates, from Eqs. (12.26) and (12.27), we get

$$v_r = 0 \qquad\qquad\qquad (12.34)$$
$$v_\theta = r\dot{\theta} = r\omega \qquad\qquad (12.35)$$

and from Eqs. (12.28) and (12.29),

$$a_r = -r\dot{\theta}^2 = -r\omega^2 \qquad (12.36)$$
$$a_\theta = r\ddot{\theta} = r\alpha \qquad\qquad (12.37)$$

The negative sign for the acceleration a_r [Eq. (12.36)] indicates that this acceleration is directed toward the center of rotation. This radial component of acceleration is known as the *centripetal acceleration*. The angular component a_θ is given by Eq. (12.37). All the above results are summarized in Fig. 12.16.

Example 12.7   Determine the acceleration toward the center of the earth and relative to the center of the earth of an object on the earth's surface at the equator. The diameter of the earth is 12.742 Mm.

Solution.   From Eq. (12.36),

$$a_r = -r\omega^2$$

$$= -6.371 \times 10^6 \left[\frac{2\pi}{24(60)(60)}\right]^2$$

$$= -3.37 \times 10^{-2} \text{ m/s}^2$$

$$= -33.7 \text{ mm/s}^2$$

Example 12.8   A belt drive is shown in Fig. 12.17. If the rotational frequency (number of revolutions per unit time) of pulley A is 5 s^{-1}, what is the rotational frequency of pulley B if the belt does not slip?

Solution.   The velocity of the belt v is equal to the velocity of a point on the circumference of either pulley. Thus,

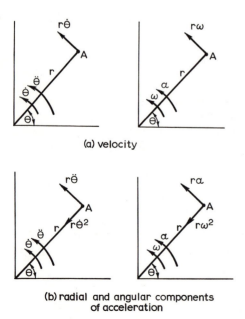

(a) velocity

(b) radial and angular components
of acceleration

Fig. 12.16   Velocity and acceleration of a point moving in a circular path of radius r.

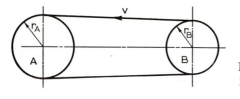

Fig. 12.17   Arrangement for Example 12.8.

$$v = \omega_A r_A = 2\pi n_A r_A \qquad \text{(a)}$$

and

$$v = \omega_B r_B = 2\pi n_B r_B \qquad \text{(b)}$$

where n_A and n_B are the rotational frequencies of pulleys A and B, respectively. Dividing Eq. (b) by Eq. (a) and rearranging, we get

$$n_B = \frac{n_A r_A}{r_B} = 5\left(\frac{r_A}{r_B}\right)$$

Problems

12.1   A car accelerates from rest with a constant acceleration of ^{5} m/s². After reaching a velocity of 10 m/s, the driver brakes, giving the car a constant deceleration and bringing the car to rest in a further 1 s. Determine the total distance traveled by the car.

12.2   The displacement of a particle is given by $s = t(t^2 + 2)$, where s is the displacement and t is the time. Plot the velocity and acceleration as a function of time for the period 0–10 s using 1-s increments.

12.3   In the October 1977 issue of *Road and Track*, the performance curves for a new SAAB turbo car were published. The following speeds for various elapsed times from a standing start were read off the published graph. Estimate the time taken for the car to travel $\frac{1}{4}$ mi from a standing start.

Elapsed Time (s)	Speed (mph)	Elapsed Time (s)	Speed (mph)
0	0	11	65
1	10	12	68
2	19	13	71
3	27	14	73
4	34	15	75
5	40	16	77
6	46	17	79
7	51	18	81
8	56	19	83
9	60	20	85
10	63		

12.4   The acceleration of a vehicle is given by $A - Bv^2$, where A and B are constants and v is the velocity of the vehicle. Obtain expressions for (a) the maximum velocity of the vehicle, (b) the time taken for the vehicle to reach half of its maximum velocity when starting from rest, and (c) the distance traveled in reaching half its maximum velocity.

12.5   A local commuter train can accelerate up to a maximum velocity of 30 m/s at a constant rate of 0.5 m/s² and can brake at a constant rate of 0.25 m/s². What is the minimum time for the train to travel between two stations (a) 1 km apart and (b) 10 km apart? What is the average velocity in each case?

12.6   A subway train leaves station A; it accelerates at the rate of 1 m/s² for 5 s and then at a rate of 2 m/s² until it has reached a velocity of 20 m/s. The train maintains the same velocity until it approaches station B; brakes are then applied, and the train decelerates at a constant rate coming to a stop 6 s after the brakes are applied. The total time for the journey is 45 s. Determine the distance between the two stations.

12.7   The velocity of a particle traveling along a straight path increases gradually from zero according to the relationship $v = 2s^{1/3}$, where s is the displacement of the particle. What will be the acceleration of the particle after it has traveled 27 m? What time will have elapsed while the particle traveled this distance?

12.8   A man is driving along a straight road at 16 m/s when the engine in his car stops running, causing a constant deceleration. At the end of 10 s he is moving at 8 m/s. How far does the car travel before it stops?

12.9   A car and a truck are both traveling at a constant velocity of 15 m/s; the car is 10 m behind the truck. The truck driver suddenly applies the brakes, causing the truck to decelerate at the constant rate of 3 m/s². Two seconds later, the driver of the car applies the brakes. Determine the constant deceleration of the car to just avoid a collision.

12.10 The angular acceleration of a body about a fixed axis is given by $\ddot{\theta} = 5 \cos \theta$. If the body starts from rest when θ is zero, what is its angular velocity after it has rotated through an angle of 30°?

12.11 For the aircraft considered in Example 12.4, assume that it is flying at a constant velocity of 100 m/s at an altitude of 2 km. Determine the rate of increase of the distance r from the tracking device to the aircraft when θ is 60°. Also determine $\ddot{r}$.

12.12 If the crank in Example 12.5 is rotating at a constant angular velocity of 6.28 rad/s, determine the radial and angular components of velocity and acceleration of the pin when θ is 60°. The radius r to the pin is 40 mm.

12.13 Figure P12.13 shows a flat-faced, spring-loaded follower being driven by an eccentric circular cam. Obtain an expression for the velocity and acceleration of the follower in terms of the constant angular velocity ω of the cam.

12.14 Figure P12.14 shows a stationary eccentric circular cam with a rotating, spring-loaded follower. Obtain expressions for the magnitude of the velocity, in polar coordinates, of the follower in terms of its constant angular velocity ω.

12.15 The loop ride shown in Fig. P12.15 malfunctioned, resulting in the small car not reaching the end of the ride. To retrieve the car, a power winch was used. The winch consisted of a 750-mm-diameter drum rotated at a constant angular velocity of 12.57 rad/s. If the top portion of the track has a shape defined by the parabola $y = x^2/15$, where x and y are measured in metres, determine the magnitude of the absolute acceleration of the car when it reaches the point where y is 1 m. Treat the car as a particle of negligible dimensions.

12.16 The equation for the spiral path shown in Fig. P12.16 is $r = r_0 + k\theta$, where k is a constant. Point P starts at A with a velocity v_0 and moves around the spiral path with constant angular velocity. Obtain equations for

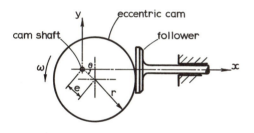

Fig. P12.13

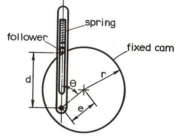

Fig. P12.14

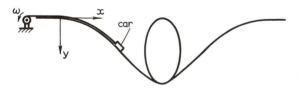

Fig. P12.15

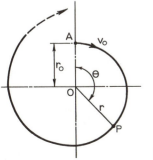

Fig. P12.16

the magnitudes of the absolute velocity and acceleration of the point P in terms of the angle θ, the initial conditions r_0 and v_0, and the constant k. If $k = 1/2\pi$, $v_0 = 1$ m/s, and $r_0 = 1$ m, what are the magnitudes of the absolute velocity and acceleration after the point has traveled one complete revolution?

12.17   A particle moves along the path described by the equation $y = -cx^2$. If v_x is a constant, show that the acceleration a_y is also a constant, and determine its magnitude in terms of c and v_x.

12.18   A train is traveling on a curved section of track of radius 1 km at a tangential velocity of 150 m/s. The brakes are suddenly applied, causing the train to slow down at a constant rate, so that after 6 s the tangential velocity has been reduced to 100 m/s. Determine the magnitude of the absolute acceleration of the train immediately after the brakes have been applied.

13

Kinematics of Rigid Bodies
(Including Applications to Mechanisms)

13.1 Introduction

A rigid body is one that does not deform under the action of forces. Thus, the distance between any two points in a rigid body is constant, and this imposes constraints on their relative motions. In this chapter we shall consider the motion of any point in a rigid body.

13.2 Plane Motion of a Rigid Body

The motion of a rigid body is said to be planar if all the points in the body undergo motion in planes that are parallel to one another. If one of these planes is the xy plane, then we may confine our attention to the cross section of the body that lies in the xy plane. All other cross sections that lie in planes parallel to the xy plane will then have identical motions relative to the x, y, z axis system.

Let us assume that the angular positions of two lines AB and CD drawn on a rigid body undergoing plane motion in the xy plane are given by θ_1 and θ_2, respectively, measured from the x axis (Fig. 13.1). If β is the angle between the lines AB and CD, measured in the xy plane, then

$$\theta_2 = \theta_1 + \beta$$

and
$$\dot{\theta}_2 = \dot{\theta}_1 \tag{13.1}$$

$$\ddot{\theta}_2 = \ddot{\theta}_1 \tag{13.2}$$

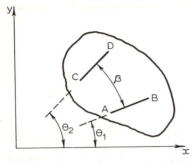

Fig. 13.1   Rigid body in plane motion.

Thus, all lines in a rigid body parallel to its plane of motion have the same angular displacement, velocity, and acceleration, and the magnitudes of these quantities coincide with those of the rigid body itself.

13.3   Velocity of a Point in a Rigid Body

Figure 13.2 shows two points, A and B, in a rigid body, separated a distance r. The instantaneous velocities of these points relative to some frame of reference are v_A and v_B, respectively. Since the body is rigid, the length r is constant, and therefore there can be no relative velocity of A to B or B to A in the direction AB. Hence, the components of v_A and v_B along AB must be equal; i.e.,

$$v_{Ax} = v_{Bx} \tag{13.3}$$

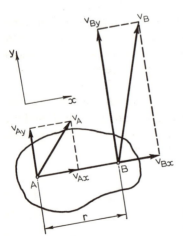

Fig. 13.2   Motion of two points in the same rigid body.

If a relative velocity between A and B exists, then it must lie in a direction perpendicular to AB, and the body will have rotary motion. Referring also to Fig. 13.3, which shows the same rotating rigid body, the angular velocity ω of the body will be given by

$$\omega = \dot{\theta} = \frac{v_{By} - v_{Ay}}{r} = \frac{v_{BA}}{r} \tag{13.4}$$

where v_{BA} is the velocity of point B relative to point A.

   If we wish to determine the absolute velocity of point B in a rigid body, this would be given by the vector addition of the absolute velocity of A and the velocity of B relative to A. Thus,

$$\mathbf{v}_B = \mathbf{v}_A + \mathbf{v}_{BA} \tag{13.5}$$

where $\mathbf{v}_{BA}$ must lie in a direction perpendicular to AB.

   In terms of the x and y components we can write

$$v_{Bx} = v_{Ax} \tag{13.6}$$

which is identical to Eq. (13.3), and

$$v_{By} = v_{Ay} + v_{BA} \tag{13.7}$$

Finally, substitution for v_{BA} from Eq. (13.4) gives

$$v_{By} = v_{Ay} + r\omega \tag{13.8}$$

These results are illustrated by the diagram in Fig. 13.4.

13.4   Acceleration of a Point in a Rigid Body

From the previous example it can be seen that when considering the relative velocity of two points in the same rigid body it can be assumed that one of the points was stationary, while the line joining the two points rotated. Applying this approach to acceleration (Fig. 13.5), we get two components of the acceleration of B relative to A as follows:

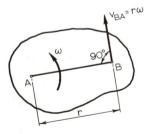

Fig. 13.3   Relative velocity of two points in a rigid body.

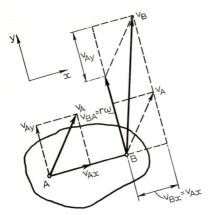

Fig. 13.4 Absolute velocity of a point in a rigid body.

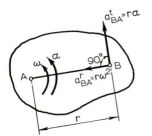

Fig. 13.5 Relative acceleration of two points in a rigid body.

1. From Eq. (12.37) and Fig. 12.16, an angular (or tangential) component (perpendicular to AB) given by

$$a_{BA}^t = r\alpha \qquad (13.9)$$

2. From Eq. (12.36) and Fig. 12.16, a radial component (parallel to AB and directed toward A) given by

$$a_{BA}^r = r\omega^2 \qquad (13.10)$$

If we wish to determine the absolute acceleration of point B in a rigid body, this is given by

$$\mathbf{a}_B = \mathbf{a}_A + \mathbf{a}_{BA}$$
$$= \mathbf{a}_A + \mathbf{a}_{BA}^t + \mathbf{a}_{BA}^r \qquad (13.11)$$

or in an xy coordinate system when AB lies in the x direction, we can write

$$a_{Bx} = a_{Ax} - r\omega^2 \qquad (13.12)$$

and $\qquad\qquad a_{By} = a_{Ay} + r\alpha \qquad (13.13)$

These results are illustrated in Fig. 13.6.

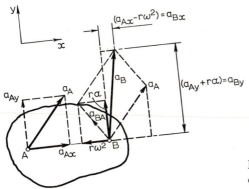

Fig. 13.6 Absolute acceleration of a point in a rigid body.

13.5 Instantaneous Center of Rotation

If a rigid body has an angular velocity, it it possible to locate a stationary point I about which the body is instantaneously rotating. This point is known as the *instantaneous center of rotation* and can be external to the body. Referring to Fig. 13.7, the velocity of point A must be normal to the line IA of length r_{IA}, and its magnitude is given by

$$v_A = r_{IA}\omega \tag{13.14}$$

Similarly, for point B,

$$v_B = r_{IB}\omega \tag{13.15}$$

It follows that if the instantaneous directions of motion of two points in a body are known, the instantaneous center of rotation for the body is located at the intersection of lines drawn from the two points perpendicular to the velocity vectors.

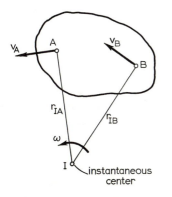

Fig. 13.7 Instantaneous center.

Example 13.1   Figure 13.8a shows a wheel which is rolling without slipping on a horizontal plane. If the velocity v_0 of the center of the wheel is 15 m/s and the radius r is 0.3 m, find the angular velocity ω of the wheel.

Solution 1.   Figure 13.8b shows the positions of the wheel before and after its center has been displaced through a distance s. The arc length $P_1'P_2$ along the rim of the wheel must be equal to s. Thus,

$$s = P_1P_2 = P_1'P_2 = r\theta \tag{a}$$

where θ is the angle through which the wheel has rotated. Differentiation of Eq. (a) with respect to time gives

$$\dot{s} = v_o = r\dot{\theta} = r\omega \tag{b}$$

Thus,

$$\omega = \frac{v_o}{r} = \frac{15}{0.3} = 50 \text{ rad/s} \quad \text{(clockwise)}$$

Solution 2.   Since the wheel is not slipping, the velocity of point P (Fig. 13.8a) is zero and is therefore the instantaneous center of rotation of the wheel. Thus, the angular velocity of the wheel is

$$\omega = \frac{v_o}{r} = \frac{15}{0.3} = 50 \text{ rad/s}$$

and is in a clockwise direction.

Example 13.2   Figure 13.9 shows a slider-crank (or simple-engine) mech-

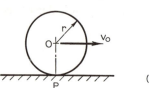

(a)

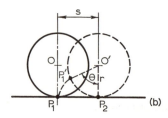

(b)   **Fig. 13.8**   Wheel rolling without slipping.

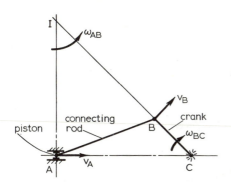

Fig. 13.9 Instantaneous center for the connecting rod of a simple-engine mechanism.

anism ABC. Find the angular velocity of the connecting rod AB and the velocity of the piston A.

Solution. Point A is constrained to move horizontally and therefore may be regarded as rotating about some point along the vertical line drawn through A. Point B, since it is attached to the crank BC, moves perpendicular to the line BC. Therefore, B may be regarded as rotating about some point on a line passing through points B and C. The only possible common point of rotation for points A and B on the connecting rod is I, which is the instantaneous center for the member AB. If the angular velocity of AB is denoted by ω_{AB}, then

$$v_B = r_{BC}\omega_{BC} = r_{IB}\omega_{AB}$$

and

$$\omega_{AB} = \frac{r_{BC}}{r_{IB}}\omega_{BC} \tag{a}$$

The velocity of the piston (point A) is given by

$$v_A = r_{IA}\omega_{AB} = \frac{r_{IA}}{r_{IB}}r_{BC}\omega_{BC}$$

or

$$v_A = \frac{r_{IA}}{r_{IB}}v_B \tag{b}$$

The dimensions r_{IA}, r_{IB}, and r_{BC} are now found by geometry, and Eqs. (a) and (b) are used to give the required values of ω_{AB} and v_A.

Example 13.3 For the simple-engine mechanism shown in Fig. 13.10a, derive equations for (a) the piston velocity, (b) the piston acceleration, and (c) the angular acceleration of the connecting rod.

Solution. If the displacement x of the piston A is measured from the position of the piston at top dead center (Fig. 13.10b), then

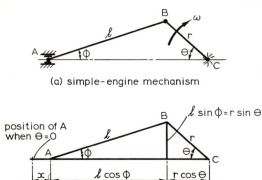

(a) simple-engine mechanism

(b) geometry

Fig. 13.10 Diagrams for Example 13.3.

$$x = r(1 - \cos \theta) + l(1 - \cos \phi) \qquad \text{(a)}$$

Now, by geometry,

$$\sin \phi = \frac{r}{l} \sin \theta \qquad \text{(b)}$$

and

$$\cos \phi = \sqrt{1 - \left(\frac{r}{l}\right)^2 \sin^2 \theta} \qquad \text{(c)}$$

Substitution for $\cos \phi$ from Eq. (c) into Eq. (a) gives

$$x = r(1 - \cos \theta) + l\left[1 - \sqrt{1 - \left(\frac{r}{l}\right)^2 \sin^2 \theta}\right] \qquad \text{(d)}$$

Differentiation of Eq. (d) gives the piston velocity v. Thus,

$$v = \dot{x} = r \sin \theta \, \dot{\theta} + \frac{l(r/l)^2 \sin 2\theta \, \dot{\theta}}{2\sqrt{1 - (r/l)^2 \sin^2 \theta}}$$

and since $\dot{\theta} = \omega$,

$$v = \omega r\left(\sin \theta + \frac{\sin 2\theta}{2\sqrt{(l/r)^2 - \sin^2 \theta}}\right) \qquad \text{(e)}$$

For most simple-engine mechanisms the ratio l/r is greater than 4, and therefore $(l/r)^2$ is greater than 16. Since the maximum value of $\sin^2 \theta$ is unity, this term can be neglected, and Eq. (e) becomes

$$v \simeq \omega r\left(\sin \theta + \frac{r}{l}\frac{\sin 2\theta}{2}\right) \qquad \text{(f)}$$

Now, differentiation of Eq. (f) gives

$$a \simeq \omega^2 r \left(\cos \theta + \frac{r}{l} \cos 2\theta \right) \qquad \text{(g)}$$

The angular velocity $\dot{\phi}$ of the connecting rod is found by differentiation of Eq. (b). Thus,

$$\cos \phi \; \dot{\phi} = \frac{r}{l} \cos \theta \; \dot{\theta}$$

or

$$\dot{\phi} = \frac{\omega r \cos \theta}{l \cos \phi} \qquad \text{(h)}$$

Substitution of Eq. (c) into Eq. (h) and rearrangement give

$$\dot{\phi} = \frac{\omega \cos \theta}{\sqrt{(l/r)^2 - \sin^2 \theta}}$$

Again, neglecting $\sin^2 \theta$ compared with $(l/r)^2$, we get

$$\dot{\phi} = \frac{r}{l} \omega \cos \theta \qquad \text{(i)}$$

Differentiation of Eq. (i) gives the angular acceleration $\ddot{\phi}$ of the connecting rod. Hence,

$$\ddot{\phi} = -\frac{r}{l} \omega \sin \theta \; \dot{\theta}$$

or

$$\ddot{\phi} = -\frac{r}{l} \omega^2 \sin \theta \qquad \text{(j)}$$

13.6 Velocity and Acceleration Diagrams

Velocity and acceleration are vector quantities, and, as with the analysis of force systems, it is often useful in the solution of problems in kinematics to draw the appropriate vector diagram.

13.6.1 Velocity Diagram

Again, taking the slider-crank or simple-engine mechanism (Fig. 13.11a) as an example, the construction of the velocity diagram (Fig. 13.11b) would be as follows. The crank is rotating with a constant angular velocity ω_{BC}, and therefore the velocity of B relative to the engine frame is

$$v_B = (BC)\omega_{BC} \qquad \text{(13.16)}$$

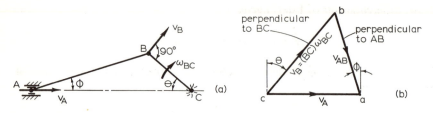

Fig. 13.11   Velocity diagram for a simple-engine mechanism.

where (BC) represents the distance between points B and C. From a point c representing the fixed point C, a line cb is drawn perpendicular to BC to represent the velocity v_B. The velocity v_A, of the piston A, must be in a horizontal direction, and therefore a horizontal line can be drawn through c. Finally, the velocity of A relative to B must be in a direction perpendicular to AB, and therefore a line is now drawn through b perpendicular to AB to intersect the horizontal line at a. The line ca now represents the velocity of the piston A, and the line ab represents the velocity of A relative to B. Now, from Eq. (13.4) the angular velocity of the connecting rod AB is given by

$$\omega_{AB} = \frac{v_{BA}}{(AB)} = \frac{(ab)}{(AB)} \qquad (13.17)$$

where (ab) represents the magnitude of v_{BA} and is the length of the line ab in the velocity diagram.

For complicated mechanisms, the velocity diagram can be drawn to scale and the distances and angles measured to give the desired magnitude and direction of the velocity of any point in the mechanism. For simple mechanisms, a sketch of the velocity diagram will allow an analysis to be made. For example, from Fig. 13.11b we can write

$$\frac{v_A}{\sin(\phi + \theta)} = \frac{v_B}{\sin[(\pi/2) - \phi]}$$

or

$$v_A = (BC)\omega_{BC}\left\{\frac{\sin(\phi + \theta)}{\sin[(\pi/2) - \phi]}\right\} \qquad (13.18)$$

Similarly,

$$v_{AB} = (BC)\omega_{BC}\left\{\frac{\sin[(\pi/2) - \theta]}{\sin[(\pi/2) - \phi]}\right\} \qquad (13.19)$$

13.6.2   Acceleration Diagram

Continuing with the same example (Fig. 13.12a), the acceleration a_{BC}^r of B relative to C is known; its magnitude is $(BC)\omega_{BC}^2$, and its direction is from B

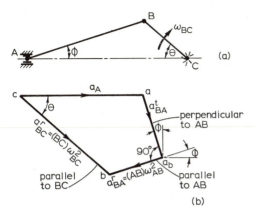

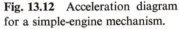

Fig. 13.12 Acceleration diagram for a simple-engine mechanism.

toward C. Thus, the line cb in the acceleration diagram (Fig. 13.12b) represents this vector. The piston A is constrained to move horizontally, and therefore point a in the acceleration diagram will lie on a horizontal line drawn through c. Considering the motion of the connecting rod AB, we note that the acceleration of B relative to A has two components: one a radial component a_{BA}^r having a magnitude $(AB)\omega_{AB}^2$ and directed toward A and the other a tangential component a_{BA}^t perpendicular to AB. Thus, a line is drawn through b parallel to AB having a length equal to $(AB)\omega_{AB}^2$, and, finally, from the end a_b of this vector, a line is drawn at right angles to AB, closing the diagram at a. The line ca now represents the magnitude of the piston acceleration a_A, and the line aa_b represents the tangential component of the acceleration of B relative to A. Again, for simple mechanisms, a sketch of the acceleration diagram can be used to obtain expressions for the acceleration of any point in the mechanism. Otherwise, the diagram must be drawn to scale.

Example 13.4   Figure 13.13a shows two sliders A and B connected by a bar 0.5 m long. The sliders are constrained to move along guides that make an angle of 30° with each other. In the position shown, slider A has a velocity v_A of 1 m/s and an acceleration a_A of 0.2 m/s² in the directions indicated. Determine (a) the angular velocity of the bar AB, (b) the angular acceleration of the bar AB, and (c) the linear acceleration of slider B.

Solution.   The velocity diagram in Fig. 13.13b is constructed as follows: (1) Assuming a stationary reference point o, the vector oa is drawn to represent v_A; (2) since point B is constrained to move horizontally, a horizontal line is drawn through o; (3) since the velocity of B relative to A must be perpendicular to AB, a vertical line is drawn through a, and its intersection with the horizontal line gives the point b.

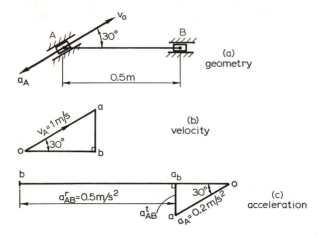

Fig. 13.13   Diagrams for Example 13.4.

Now the angular velocity of the bar AB is given by

$$\omega_{AB} = \frac{(ab)}{(AB)} = \frac{1(\sin 30°)}{0.5} = 1 \text{ rad/s} \qquad \text{(clockwise)}$$

To construct the acceleration diagram (Fig. 13.13c), it is first necessary to calculate the radial component of the acceleration of A relative to B. This is given by

$$a^r_{AB} = (AB)\omega^2_{AB} = 0.5(1)^2 = 0.5 \text{ m/s}^2$$

and is directed from A toward B. The line oa is now drawn at an angle of 30° to represent the absolute acceleration of A, which is given as 0.2 m/s². Since point B is constrained to move horizontally, a horizontal line is drawn through o. A vertical line drawn through a intersects the horizontal line at a_b, and the line aa_b represents the tangential component a^t_{AB} of the acceleration of A relative to B. Since the radial component a^r_{AB} of the acceleration of A relative to B is horizontal, this vector (length, 0.5 m/s²) must lie on the horizontal line through o. Thus, point b is determined as shown in Fig. 13.13c.

Now the angular acceleration of the bar AB is given by

$$\alpha_{AB} = \frac{(aa_b)}{(AB)} = \frac{0.2 \sin 30°}{0.5} = 0.2 \text{ rad/s}^2 \qquad \text{(counterclockwise)}$$

Finally, the linear acceleration of the slider B is given by

$$(ob) = 0.2 \cos 30° + 0.5$$
$$= 0.673 \text{ m/s}^2 \qquad \text{(to the left)}$$

Problems

13.1 The wheel of radius r, shown in Fig. P13.1, is rolling without slipping along a horizontal surface. Its angular velocity is ω (clockwise), and its angular acceleration is α (clockwise). For the position shown, find (a) the magnitudes of the vertical and horizontal components (v_y, v_x) of the velocity of point P on the wheel's circumference and (b) the magnitudes of the vertical and horizontal components (a_y, a_x) of the acceleration of point P. In each case, indicate the direction of the component by a sign.

13.2 Figure P13.2 shows the plan view of a car door AB of width $2b$ which has been given a constant angular velocity of ω rad/s in order to close it. At the same time, the car is starting to move forward from rest with an acceleration of a m/s². Find, for the position shown, (a) the absolute velocity v_P of a point P, on the car door, midway between A and B; (b) the component, in the direction of motion of the car, of the absolute acceleration a_P of the point P; and (c) the angular acceleration of the car door.

13.3 Figure P13.3 shows a 1-m-diameter disk rolling without slipping on a horizontal platform P which is moving to the right with a velocity v_P of 1 m/s. If the angular velocity of the disk is 0.5 rad/s, in a counterclockwise direction, determine the absolute velocity of the center of the disk.

13.4 Figure P13.4 shows a cable wrapped around an axle used to connect two

$$13.3 \quad V_0 = V_{0x} = 0.75 \, m/s$$

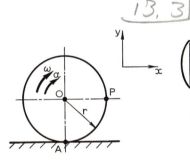

Fig. P13.1

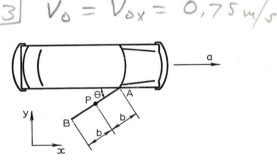

Fig. P13.2

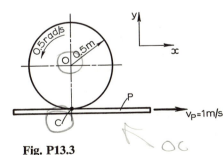

Fig. P13.3

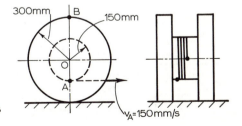

Fig. P13.4

$$WR + 1 \, m/s$$

wheels arranged side by side. If the cable is pulled to the right with a constant velocity of 150 mm/s and if the wheels roll without slipping, determine the magnitudes of the velocity and acceleration of point B on the periphery of one of the wheels.

13.5 The ladder AB (Fig. P13.5), which is 5 m long, is sliding so that the lower end B has a constant velocity, away from the vertical wall, of 1 m/s. Determine the velocity and acceleration of the upper end A of the ladder when the lower end is 3 m from the base of the wall.

13.6 A plank AB, 2 m long, is resting against a cylinder of radius 0.5 m and at an angle of 60° to the horizontal (Fig. P13.6). The end B is slid toward the cylinder at a velocity of 5 m/s while maintaining contact with the floor. (a) Determine the x and y components of the velocity of end A of the plank. (b) What will these components be when the plank becomes vertical?

13.7 Figure P13.7 shows a bar AB leaning against a step and moving in such a way that end A has a constant velocity of 3 m/s along the horizontal surface. Determine the absolute velocity of point C on the bar when the bar is inclined 30° to the horizontal.

13.8 A car with 700-mm-diameter tires accelerates from rest to 30 m/s in a distance of 400 m. If the acceleration of the car is constant, determine the

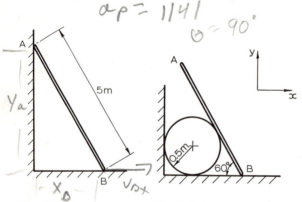

Fig. P13.5   Fig. P13.6

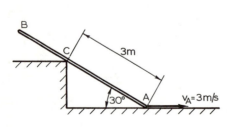

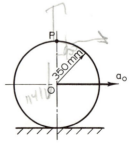

Fig. P13.7   Fig. P13.8

$$\frac{d\,v_{ay}}{dt} = -\frac{v_b{}^2 v_{bx}{}^2}{(25-x_a^2)^{3/2}} - \frac{v_{bx}{}^2}{(25-x_a^2)^{1/2}} = -0.39$$

magnitude of the acceleration of point P on the periphery of the tire (Fig. P13.8) when the car reaches a speed of 20 m/s.

13.9 A 30° wedge slides along a horizontal plane at 10 m/s while a block slides up the wedge at a velocity, relative to the wedge surface, of 5 m/s (Fig. P13.9). What is the absolute velocity of the block?

13.10 The arm AB of a planetary gear assembly (Fig. P13.10) rotates at a constant angular velocity ω. Determine the x, y components of velocity and acceleration of point P on the periphery of the planetary gear.

13.11 Determine the x, y components of velocity and acceleration of point P on the periphery of a wheel of radius r_w which is rolling without slipping on the inside of a fixed circular track of radius r_t (Fig. P13.11). The arm connecting the fixed point A and the center of the wheel B is rotating with a constant angular velocity ω.

13.12 In the four-bar mechanism shown in Fig. P13.12 the crank AB rotates at a frequency of $1\ \mathrm{s}^{-1}$ in a clockwise direction. Locate the instantaneous

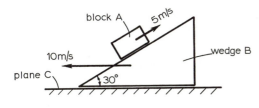

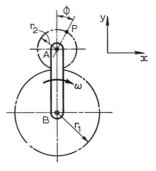

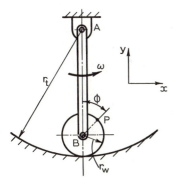

Fig. P13.9

Fig. P13.10

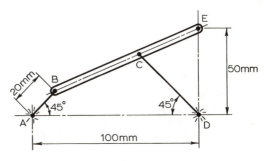

Fig. P13.11

Fig. P13.12

center of rotation for the rigid link *BCE*, and hence determine the magnitude and direction of the instantaneous velocity of point *E*.

13.13   A slender rod *AB* connects two cranks *C* and *D* as shown in Fig. P13.13. If crank *C* is rotating counterclockwise at an angular velocity of 2 rad/s, determine the angular velocity of crank *D* for the instant shown.

13.14   In the simple-engine mechanism shown in Fig. P13.14, the crank *AB* rotates at a constant angular velocity of 20 rad/s. Locate the instantaneous center of rotation of the connecting rod *BC*, and hence determine the velocity of the piston *C* and the angular velocity of the connecting rod *BC*.

13.15   In the quick-return mechanism shown in Fig. P13.15, the crank *CA* rotates at a constant angular velocity ω. Derive expressions for the absolute velocity and acceleration of the slider *B* in terms of the angular position of the crank θ and the crank radius *r*.

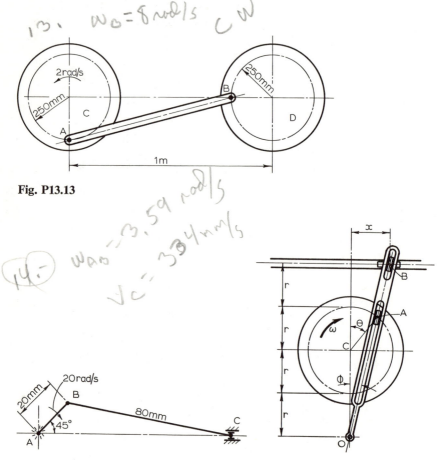

Fig. P13.13

Fig. P13.14

Fig. P13.15

13.16 The slotted link AB shown in Fig. P13.16 is caused to oscillate about A by
the action of the crank pin C, which is free to slide in the slot. If the crank
rotates at a constant angular velocity $\dot\theta$ as shown, derive equations for the
angular velocity and acceleration $\dot\phi$ and $\ddot\phi$, respectively, of the link AB.
If $l/r = 2$, $\theta = 60°$, and $\dot\theta = 1$ rad/s, determine the magnitudes of $\dot\phi$ and $\ddot\phi$.
Check your answer by drawing the velocity and acceleration diagrams to
scale.

13.17 For the four-bar mechanism shown in Fig. P13.17, sketch carefully the
velocity and acceleration diagrams, and hence determine the instantaneous
angular velocity and acceleration of the link CD.

13.18 Figure P13.18 shows a four-slotted Geneva wheel which, when the crank
is rotated at constant speed, is provided with intermittent rotary motion.
For smooth action, the angular velocity $\dot\phi$ of the wheel must be zero as the
pin on the crank enters or leaves one of the slots. If both the radius to the

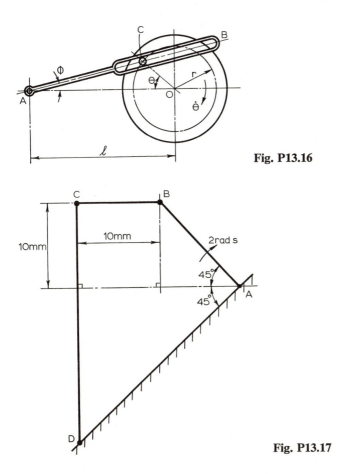

Fig. P13.16

Fig. P13.17

center of the pin and the radius of the slotted wheel are equal to r, determine the distance b between the two centers to give smooth action. In addition, develop an expression for the angular acceleration α_w of the wheel in terms of the constant angular velocity ω_c of the crank during indexing.

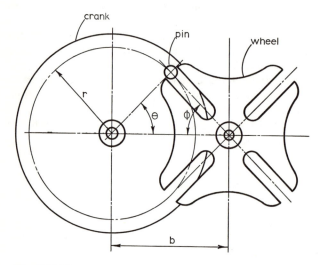

Fig. P13.18

14

Dynamics of Particles

14.1 Introduction

In the previous chapters on kinematics our attention was devoted entirely to a study of the relationship between the fundamental quantities of length and time and the derived quantities of velocity and acceleration. Our concern there was a study of the motion of particles and rigid bodies without regard to the forces causing the motion. In this chapter the analysis of the motion of a particle as a result of the external forces acting on it is considered. In addition, for those cases where the motion of the particle is specified, the forces required to produce this motion will be determined. It will be assumed that rotational effects can be neglected and that the body can be treated as a single particle with its mass concentrated at a point. If rotational effects cannot be neglected, then the problem must be treated by the more general methods of rigid body dynamics dealt with in later chapters. It should be noted that the same body might in one case be treated as a particle, while in another case it might be treated as a rigid body. For example, a communications satellite orbiting the earth can be treated as a particle for the purposes of determining its orbit about the earth; this same satellite would be treated as a rigid body if, for the purposes of communication, its orientation relative to the surface of the earth is important.

14.2   Equations of Motion

Newton's second law of motion can be stated as follows: "A particle acted upon by a force will move with an acceleration proportional to and in the direction of the force." Since the acceleration is always in the direction of the force, the relationship is really a vector relationship and should be written as follows:

$$\mathbf{F} = m\mathbf{a} \tag{14.1}$$

This equation holds only for a particle with constant mass and means that the vector representing the resultant force acting on the particle is equal in magnitude and direction to the vector representing its acceleration multiplied by the mass m. Equation (14.1) is often referred to as the equation of motion.

Since Eq. (14.1) is a vector equation, its scalar components may be written for any of the various coordinate systems. Thus, for rectangular coordinates, the component equations are

$$F_x = ma_x = m\ddot{x} \tag{14.2}$$

and
$$F_y = ma_y = m\ddot{y} \tag{14.3}$$

The importance of free-body diagrams in obtaining the equations of motion for a given mechanical system should not be underestimated. As in statics, experience has shown that the vast majority of errors made in dynamics can be traced to either the lack of a free-body diagram or an incorrect free-body diagram.

Figure 14.1a shows a body of mass m on a frictionless horizontal surface; it is being pulled with a force F_p by means of a string. Figure 14.1b shows the external forces acting on the body; this is a free-body diagram. Figure 14.1c shows the acceleration $\ddot{x}$ of the mass multiplied by the mass m. According to Eq. (14.2), the product $m\ddot{x}$, which represents the resistance of the body to acceleration and is often referred to as the inertia force, can be equated to the net horizontal component F_x of all the external forces shown in Fig. 14.1b. Thus,

$$F_x = F_p \cos \theta = m\ddot{x}$$

or
$$\ddot{x} = \frac{F_p}{m} \cos \theta \tag{14.4}$$

and knowing the values of F_p, θ, and m, the resulting acceleration can be found. Figure 14.1c can be regarded as an inertia-force diagram since it shows the inertia forces acting on the body.

In our subsequent work we shall adopt the following procedure when equations of motion are required:

1. Draw a free-body diagram showing the external forces acting.

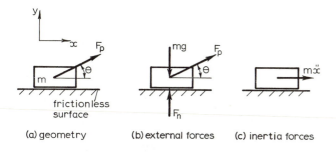

(a) geometry   (b) external forces   (c) inertia forces

Fig. 14.1 Illustration of force diagrams.

2. Draw an inertia-force diagram showing the inertia forces.
3. Equate the appropriate components of the forces shown in the two diagrams to obtain the required component equations of motion.

In many problems, once the component equations of motion have been written for the given situation, the solution to the equations must be obtained. In some simplified situations direct integration may be used; in other cases numerical or graphical means are required. The difficulties encountered depend on the types of forces present, that is, whether the force is a constant, such as weight, or whether the force is a variable, such as the "drag" of a fluid on a solid body.

Example 14.1 What horizontal force F_p is required to accelerate a mass of 10 kg at 20 m/s² across a horizontal surface if the coefficient of friction between the mass and the surface is 0.3?

Solution. Figure 14.2a shows the free-body diagram for the mass, where mg is the weight of the mass, F_n is the reaction provided by the horizontal surface, and F_f is the frictional force. Figure 14.2b shows the inertia force due to the acceleration of the mass. Equating the horizontal components of the forces shown in Figs. 14.2a and 14.2b, we get

$$F_p - F_f = m\ddot{x} \qquad \text{(a)}$$

Since the body is sliding,

$$F_f = \mu F_n \qquad \text{(b)}$$

and substitution of Eq. (b) into Eq. (a) gives

$$F_p - \mu F_n = m\ddot{x} \qquad \text{(c)}$$

Equating the vertical components of the forces in Figs. 14.2a and 14.2b, we get

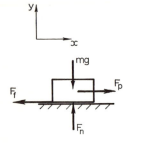
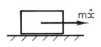

(a) external forces   (b) inertia forces   **Fig. 14.2** Diagrams for Example 14.1.

$$F_n - mg = 0$$

or $$F_n = mg \qquad \text{(d)}$$

Substitution of Eq. (d) into Eq. (c) gives

$$F_p - \mu mg = m\ddot{x}$$

or $$F_p = m(\mu g + \ddot{x}) \qquad \text{(e)}$$

Since m is 10 kg, $\ddot{x}$ is 20 m/s², and μ is 0.3, we get

$$F_p = 10[0.3(9.81) + 20] = 229 \text{ N}$$

Example 14.2 A shuffleboard disk is sent sliding down the court with an initial velocity of 5 m/s. If the coefficient of friction μ between the disk and the court is 0.2 how far will the disk travel before it comes to rest?

Solution. The free-body and inertia-force diagrams for the disk as it slides down the court are shown in Fig. 14.3; mg is the weight of the disk, F_n is the normal reaction between the court and the disk, and μF_n is the frictional force. The distance x will be measured along the surface of the court from the point where the disk is released.

Equating the horizontal components of the forces shown in Fig. 14.3 yields

$$-\mu F_n = m\ddot{x} \qquad \text{(a)}$$

Also, by equating the vertical components, F_n is equal to mg, and Eq. (a) becomes

$$\ddot{x} = -\mu g \qquad \text{(b)}$$

Now, $\ddot{x}$ can be written as dv/dt, and therefore, from Eq. (b),

$$dv = -\mu g \, dt \qquad \text{(c)}$$

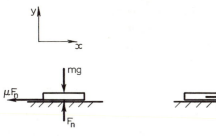

Fig. 14.3 Diagrams for Example 14.2.

(a) external forces   (b) inertia forces

Integration of Eq. (c) gives

$$v = -\mu g t + C_1 \qquad\qquad (d)$$

where C_1 is a constant of integration. Since v is 5 m/s when t is zero, C_1 is 5, and Eq. (d) becomes

$$v = -\mu g t + 5 \qquad\qquad (e)$$

Since v can be written as dx/dt, Eq. (e) becomes

$$dx = (-\mu g t + 5)\, dt \qquad\qquad (f)$$

and integration of Eq. (f) gives

$$x = \frac{-\mu g t^2}{2} + 5t + C_2 \qquad\qquad (g)$$

Since x is zero when t is zero, C_2 is zero, and Eq. (g) becomes

$$x = \frac{-\mu g t^2}{2} + 5t \qquad\qquad (h)$$

When the disk comes to rest its velocity is zero, and, from Eq. (e), this occurs when

$$t = \frac{5}{\mu g} \qquad\qquad (i)$$

Substitution of Eq. (i) into Eq. (h) gives

$$x = \frac{-25}{2\mu g} + \frac{25}{\mu g} = \frac{25}{2\mu g}$$

or

$$x = \frac{25}{2(0.2)(9.81)} = 6.37 \text{ m}$$

Example 14.3 A parachutist of mass 90 kg jumps from a helicopter hovering at a high altitude and falls under the action of gravity for 10 s, at which time

he opens his parachute. Develop an equation for the velocity of the parachutist after the parachute opens. What is the minimum velocity with which
the parachutist can hit the ground? Assume that air resistance is negligible
until the parachute opens, when the air resistance is given by kv^2, where k is a
constant and is equal to $50 \text{ N} \cdot \text{s}^2/\text{m}^2$.

Solution. For this problem, the x distance will be measured from the helicopter toward the ground. Since, during free-fall, air resistance is neglected,
the free-body and inertia-force diagrams for the parachutist of weight mg are
as shown in Fig. 14.4a. Equating the vertical components of the forces on the
parachutist gives

$$mg = m\ddot{x}$$

or $$\ddot{x} = g \qquad (a)$$

Integrating twice and writing $x = \dot{x} = 0$ when $t = 0$, we get

$$\dot{x} = gt \qquad (b)$$

$$x = \frac{gt^2}{2} \qquad (c)$$

The parachutist's velocity v_o and displacement x_o, after 10 s, are obtained by
substituting $t = 10$ into Eqs. (b) and (c); thus,

$$v_o = \dot{x} = 98.1 \text{ m/s} \qquad (d)$$

$$x_o = x = 490.5 \text{ m} \qquad (e)$$

These values of v_o and x_o become the initial conditions for the equation of
motion for the parachutist once his chute opens.

The free-body and inertia-force diagrams for the parachutist once his chute
opens are shown in Fig. 14.4b. Here, it is assumed that the air resistance
acting on the chute is proportional to the square of the instantaneous speed

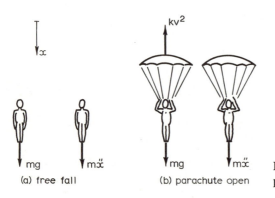

(a) free fall   (b) parachute open

Fig. 14.4   Diagrams for Example 14.3.

and acts in a direction opposite to the velocity vector. Now, equating the forces in the two diagrams, we get

$$m\ddot{x} = m\frac{dv}{dt} = -kv^2 + mg \tag{f}$$

Integration of Eq. (f) gives

$$\int \frac{dv}{(mg/k) - v^2} = \int \frac{k}{m} dt$$

and therefore,

$$\frac{1}{2\sqrt{mg/k}} \ln \left(\frac{\sqrt{mg/k} + v}{\sqrt{mg/k} - v}\right) = \frac{k}{m}t + C \tag{g}$$

Since $v = v_o$ when $t = 0$,

$$C = \frac{1}{2\sqrt{mg/k}} \ln \left(\frac{\sqrt{mg/k} + v_o}{\sqrt{mg/k} - v_o}\right)$$

Thus,

$$\frac{\sqrt{mg/k} - v}{\sqrt{mg/k} + v} = \left(\frac{\sqrt{mg/k} - v_o}{\sqrt{mg/k} + v_o}\right) \exp\left(-2\sqrt{\frac{gk}{m}}t\right) \tag{h}$$

Substitution of the values of m, g, k, and v_o gives

$$\frac{4.2 - v}{4.2 + v} = -0.9179e^{-4.67t} \tag{i}$$

and, after rearrangement we get

$$v = 4.2\left(\frac{1 + 0.92e^{-4.67t}}{1 - 0.9179e^{-4.67t}}\right) \tag{j}$$

As t increases, the right-hand side of Eq. (h) rapidly approaches zero, and

$$v = \sqrt{\frac{mg}{k}} = 4.2 \text{ m/s} \quad (9.4 \text{ mph}) \tag{k}$$

Equation (k) could have been immediately obtained from Eq. (f) by realizing that when the velocity becomes constant, the acceleration is zero.

Example 14.4 A small spherical particle is placed on top of a cylinder of radius r which is held with its axis horizontal. If the particle is given a small displacement, what angle will it subtend from the vertical when it leaves the surface of the cylinder.

Solution. Figure 14.5 shows the free-body and inertia-force diagrams for the particle when it has rolled through an angle θ. The reaction normal to the

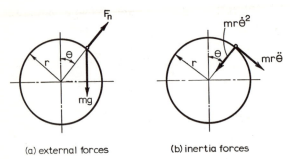

(a) external forces   (b) inertia forces

Fig. 14.5 Diagrams for Example 14.4.

surface is F_n. Resolving in a direction tangential to the surface of the cylinder, we get

$$mg \sin \theta = mr\ddot{\theta} \qquad\qquad\text{(a)}$$

Resolving radially, we get

$$mg \cos \theta - F_n = mr\dot{\theta}^2 \qquad\qquad\text{(b)}$$

From Eq. (a),

$$g \sin \theta = r\dot{\theta}\frac{d\dot{\theta}}{d\theta}$$

or

$$\frac{g}{r} \int \sin \theta \, d\theta = \int \dot{\theta} \, d\dot{\theta}$$

and, by integration,

$$-\frac{g}{r} \cos \theta = \frac{\dot{\theta}^2}{2} + C \qquad\qquad\text{(c)}$$

When $\theta = 0$, $\dot{\theta} = 0$, and $C = -(g/r)$. Thus,

$$\frac{\dot{\theta}^2}{2} = \frac{g}{r}(1 - \cos \theta) \qquad\qquad\text{(d)}$$

Eliminating $\dot{\theta}$ from Eqs. (b) and (d) and realizing that the particle leaves the surface of the cylinder when $F_n = 0$, we get

$$\frac{mg}{2r} \cos \theta = \frac{mg}{r}(1 - \cos \theta) \qquad\qquad\text{(e)}$$

and finally, solving for θ, we get

$$\tfrac{1}{2} \cos \theta = 1 - \cos \theta$$

$$\cos \theta = \tfrac{2}{3}$$

$$\theta = 48.19°$$

Example 14.5   A kicker in a football game wishes to kick the ball a maximum distance. At what angle, measured from the horizontal, should he kick the ball if air resistance can be neglected?

Solution.   The trajectory of the football and the free-body and inertia-force diagrams are shown in Fig. 14.6. The component equations of motion are

$$m\ddot{x} = 0 \qquad\qquad (a)$$

and

$$m\ddot{y} = -mg \qquad\qquad (b)$$

Hence, from Eq. (a),

$$\ddot{x} = 0$$

or, by integration,

$$\dot{x} = C \qquad\qquad (c)$$

When $t = 0$, $\dot{x} = v_o \cos \theta$, and therefore,

$$\dot{x} = v_x = v_o \cos \theta \qquad\qquad (d)$$

where v_o is the initial absolute velocity of the ball and θ is the angle at which the ball is kicked. If the total horizontal distance traveled by the ball is denoted by s, then the total time of flight t_f is given by

$$t_f = \frac{s}{v_x} = \frac{s}{v_o \cos \theta} \qquad\qquad (e)$$

From Eq. (b) we get

$$\ddot{y} = -g$$

or

$$\frac{dv_y}{dt} = -g \qquad\qquad (f)$$

Integration of Eq. (f) twice and writing $v_y = v_o \sin \theta$ and $y = 0$ when $t = 0$,

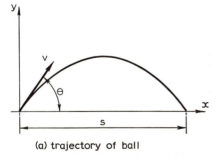

(a) trajectory of ball   (b) external force   (c) inertia force

Fig. 14.6 Diagrams for Example 14.5.

we get

$$y = v_o t \sin \theta - \frac{gt^2}{2} \tag{g}$$

When the ball falls to the ground, $y = 0$, and thus, from Eq. (g), the time taken is given by

$$t_f = \frac{2v_o \sin \theta}{g} \tag{h}$$

Equating Eqs. (e) and (h) and rearranging, we get

$$s = \frac{2v_o^2}{g} \sin \theta \cos \theta \tag{i}$$

To find the value of θ that will maximize the horizontal distance traveled, Eq. (i) can be differentiated with respect to θ and the result equated to zero. Thus,

$$\frac{ds}{d\theta} = \frac{2v^2}{g}(-\sin^2 \theta + \cos^2 \theta) = 0 \tag{j}$$

or $\tan \theta = 1$

and $\theta = 45°$

Problems

14.1   The effect of velocity on the total air resistance to the horizontal motion of a 1000-kg rocket test sled can be approximated by the straight-line graph shown in Fig. P14.1. If, when the rocket burns out, the velocity of the test sled is 250 m/s, determine the further distance the test sled travels before coming to rest.

14.2   A man standing on a bridge, 20 m above the water, throws a stone in a direction 30° above the horizontal. Knowing that the stone hits the water

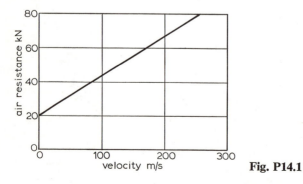

Fig. P14.1

30 m horizontally from a point below the man, determine (a) the initial velocity of the stone and (b) the distance at which the stone would hit the water if it were thrown in the same way from a point 3 m lower.

14.3 Using the data in Example 14.3, plot a graph of velocity on a base of time from the instant when the parachutist leaps from the helicopter to 1.0 s after the parachute opens. Use increments of 0.1 s after the chute opens. From these results, estimate how long it takes for the parachutist to reach the ground from the helicopter, which is hovering at an altitude of 1 km.

14.4 A particle of mass m, moving in the xy plane, is attracted toward the x axis by a force $\alpha m/y^3$ when at a point (x, y). Show that if α is constant and the particle is projected from the point $(0, k)$ with commponent velocities v_x and v_y parallel to the axes x and y, respectively, y will achieve a maximum value if $\alpha > k^2 v_y^2$.

14.5 Find the minimum velocity with which a particle of mass m must be projected radially away from the earth's surface so that it escapes from the effect of the earth's gravitational pull. Neglect the effects of other planets. The universal gravitational constant G is 6.673×10^{-11} m³/kgs², the radius of the earth r_e is 6.371×10^6 m, and the mass of the earth m_e is 5.976×10^{24} kg.

14.6 The pendulum in Fig. P14.6 is held in the position shown. Determine the ratio of the tensions in the wire AB immediately before and after cutting the wire CD.

14.7 For the arrangement shown in Fig. P14.7, determine its minimum angular velocity so that the two 60-N spheres do not contact the conical surface.

14.8 A mass m connected to the weightless thread of length l is held in the position shown in Fig. P14.8 and then released. Obtain expressions for (a) the velocity of the mass when it reaches its lowest position and (b) the tension in the thread when the mass reaches its lowest position.

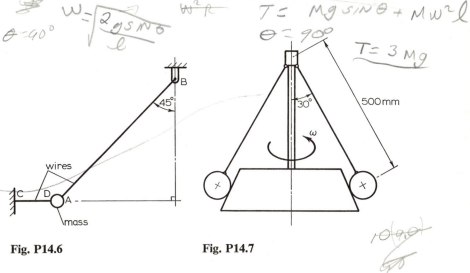

Fig. P14.6   **Fig. P14.7**

$14.11) \quad 706 \ \text{\~{A}}$

$14.12) \quad W_0 \mp \sqrt{\dfrac{g}{R}(3\cos\theta - 2)}$

$\theta = 30°$

$\sqrt{0} = W_0 R = 7.66$

14.9 A racing cyclist is traveling at a constant speed of 7 m/s around the elliptical track shown in Fig. P14.9. If the center of mass of the cyclist is 1.2 m above the ground, what is the maximum angle from the vertical through which the cyclist must lean?

14.10 A spring scale, suspended from the ceiling of an airplane, is used to weigh a package while the airplane is making a 1-km-radius turn at 150 m/s. If the scale reads 40 N, what is the true weight of the package? Neglect the weight of the moving parts of the scale.

14.11 An amusement park ride involves a 3-kN car which travels on a track. The car is traveling at a constant speed of 15 m/s as it passes over a vertical hump in the track. If the radius at the top of the hump is 30 m, determine the normal reaction between the track and the car. Determine this same reaction when the car negotiates a dip of the same curvature.

14.12 A roller-coaster track consists of circular arcs connected tangentially with straight portions of track, all in a vertical plane. A car traveling on the level commences to descend on a circular portion having a radius of 10 m in a vertical plane. The maximum gradient on this portion is 30° to the horizontal. Neglecting friction, what is the car's maximum initial velocity if the car is not to leave the track?

14.13 If the coefficient of friction between the 1-kg mass and the plane shown in Fig. P14.13 is 0.2, determine how long it will take for the mass to slide 0.2 m up the plane, starting from rest.

14.14 A particle of mass m_1 is placed on the smooth face of an inclined plane of mass m_2 and slope α. The inclined plane is free to move on a smooth horizontal plane. Show that if there is no friction and the system starts from rest, the ratio of the velocity attained by the inclined plane to the velocity of the particle relative to the plane in a given time interval is $(m_1 \cos \alpha)/(m_1 + m_2)$. If m_1 is 1 kg, m_2 is 2 kg, and α is 30°, find the velocity of the particle relative to the plane after 2 s.

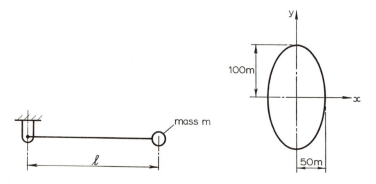

Fig. P14.8   Fig. P14.9

14.15 A block of mass m is initially at rest on a plane which is inclined at an angle α to the horizontal. A horizontal force F_p is applied to the block so as to push it up the plane a distance s. If the coefficient of friction between the block and the plane is μ, obtain an expression for the final velocity of the block.

14.16 A 5-kg block is released from the top of a 45° inclined plane. If the coefficient of sliding friction between the plane and the block is 0.2, how long will it take for the block to slide 0.5 m down the plane? What will the velocity of the block be after this time?

14.17 A boy is whirling a ball which is attached to a weightless string. The motion of the ball is in a vertical plane, and its weight is 1 N. If the length of the string is 1.5 m and the ball completes one revolution every 1.2 s at constant speed, what is the magnitude of the maximum tension in the string each revolution?

$W = ?$

$\dfrac{2\pi}{1.2}$

$= 5,236 \dfrac{rad}{s}$

$T_{MAX} = MW^2 R + F_W = 5,19 N$

14.18 A particle of mass m_1 is placed on the smooth face of an inclined plane of mass m_2 and slope α. The inclined plane is free to move on a smooth horizontal plane. The plane is pushed with a force F_p so that the particle remains stationary on the plane. Obtain an expression for the force F_p.

14.19 A mass m lies at rest on a horizontal table and is attached to one end of a massless spring which, when stretched, exerts a tension of magnitude mw^2 times its extension (where w is a constant). If the other end of the spring is now moved with uniform velocity v along the table in a direction away from the mass and the table offers a resistance to the motion of the mass of magnitude mk times its speed (where k is constant), obtain the differential equation for the extension x of the spring after time t.

14.20 Shown in Fig. P14.20 are three blocks A, B, and C having masses of 5 kg, 10 kg, and 30 kg, respectively. The blocks are connected to each other by by inextensible strings. The coefficient of friction between the blocks and the surfaces is 0.2. Neglecting the friction in the pulleys, determine the acceleration of the blocks and the tensions in the two strings AB and BC.

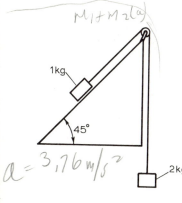

$M_1 + M_2(a)$

1kg

45°

$a = 3,76 m/s^2$

2kg

Fig. P14.13

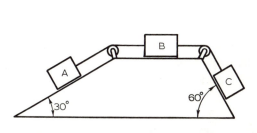

A

B

C

30°

60°

Fig. P14.20

$v = 3,76 t$

$x = 1,88 t^2$

$x = a\dfrac{2}{t} = 1,33 s$

15

Work, Power, and Energy Applied to the Dynamics of Particles

15.1 Introduction

In the previous chapter the velocity and position of a particle were obtained by direct integration of the equations of motion with respect to time. For certain classes of problems, however, it is best to obtain the velocity of a particle by use of the work-energy equation, which is the first integral of the equation of motion with respect to displacement. The derivation of the work-energy equation and its use are the subjects of this chapter.

15.2 Work-Energy Equation

If a constant force F acts on a particle so as to produce a displacement s of the particle in the direction of the force, the *work* W is defined as the product of the force and the displacement; i.e.,

$$W = Fs \tag{15.1}$$

If the magnitude of the force changes continuously during the displacement, the work done during any small displacement Δs is given by $F \Delta s$, and the total work done during a finite displacement is given by

$$W = \int F \, ds \tag{15.2}$$

Since work is the product of a force and a length, it has the unit of newton-metre (N·m), but this unit is given the special name of joule (J). One joule is the work done by a force of 1 N through a distance of 1 m.

Power is the rate of doing work, or the work done in unit time. Thus, power P is given by

$$P = \dot{W} = F\frac{ds}{dt} = Fv \tag{15.3}$$

where v is the velocity of the particle. The unit of power is newton-metre per second (N·m/s) or joule per second (J/s) and is given the special name watt (W).

15.2.1 Rectilinear Motion

For a single particle moving along a straight line, the equation of motion for the particle in the direction of motion is given by

$$F = m\frac{dv}{dt} = m\frac{dv}{ds}\frac{ds}{dt} = mv\frac{dv}{ds} \tag{15.4}$$

where s is the distance from some reference point on the line and F is the force acting in the s direction. If the particle moves from point A to point B, then substitution of Eq. (15.4) into Eq. (15.2) and insertion of the limits A and B give

$$W = m \int_{v_A}^{v_B} v \, dv$$
$$= \tfrac{1}{2}mv_B^2 - \tfrac{1}{2}mv_A^2 \tag{15.5}$$

Equation (15.5) allows us to calculate the change in velocity of the particle due to the application of a force. If the force and displacement are in the same direction, then the work W is positive, and the velocity of the particle increases. If the force and displacement are in opposite directions, then the work W is negative, and the velocity of the particle decreases.

At this point we can introduce the concept of energy. *Energy* is sometimes defined as the capacity for doing work. There are many different forms of energy, such as heat, electric, potential, kinetic, etc. The mechanical forms of energy are *potential* and *kinetic* energy. Kinetic energy is the energy due to motion. The terms $\tfrac{1}{2}mv_A^2$ and $\tfrac{1}{2}mv_B^2$ in Eq. (15.5) represent the kinetic energy of the particle at points A and B, respectively. Thus, if we denote kinetic energy by the symbol T, we can write

$$T = \tfrac{1}{2}mv^2 \tag{15.6}$$

where m is the mass and v is the velocity of the particle.

The amount of kinetic energy possessed by a particle may be regarded as the work necessary to bring the particle to rest and is given by Eq. (15.6).

Substitution of Eq. (15.6) into Eq. (15.5) gives

$$W = T_B - T_A = \Delta T \qquad (15.7)$$

This equation is called the *work-energy equation* and means that as the particle moves from A to B, the work done on the particle by external forces applied in the direction of motion is equal to the change in its kinetic energy.

Example 15.1   In Example 14.2, a shuffleboard disk was propelled at an initial velocity of 5 m/s and it was desired to calculate how far the disk would travel before coming to rest if the coefficient of friction were 0.2. This problem can now be solved using the work-energy equation.

Solution.   From the diagram in Fig. 15.1, the frictional force opposing motion is constant and equal to μmg. If the total distance traveled by the disk before coming to rest is s, then the work done by the frictional force F_f is given by

$$W = -\mu mgs \qquad (a)$$

where the minus sign indicates that the direction of the force is opposite to that of the displacement. It should be noted that since frictional forces always oppose motion, the work done by these forces will always be negative in Eq. (15.7).

The initial velocity v_A of the disk is 5 m/s, and the final velocity is zero. Thus, T_A is $\frac{1}{2}m(5)^2$, and T_B is zero. From Eq. (15.7),

$$W = T_B - T_A \qquad (b)$$

and substitution for W from Eq. (a) gives

$$-\mu mgs = -\tfrac{1}{2}m(5)^2$$

$$s = \frac{25}{2\mu g} = \frac{25}{2(0.2)(9.81)} = 6.37 \text{ m}$$

Example 15.2   An elevator is used to raise a load of furniture from the basement to the third floor of a building, a distance of 30 m, in 10 s. The total

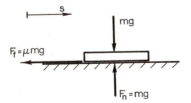

Fig. 15.1   Free-body diagram for Example 15.1.

mass of the elevator, operator, and furniture is 2 Mg. What is the average power required?

Solution. Power is the rate of doing work and is given by Eq. (15.3). Thus,

$$P = Fv$$

$$= 2000(9.81)\left(\frac{30}{10}\right) = 58.9 \text{ kW}$$

15.2.2 Curvilinear Motion

Until now, we have only considered a particle moving in a straight path acted upon by a force applied parallel to the direction of motion. We shall now show that the work-energy equation (15.7) also applies to curvilinear motion. Figure 15.2 shows a particle moving along a curved path and acted upon by a force which, at a given instant, has a magnitude F and makes an angle α with the tangent to the path of the particle. The equation of motion at point P for the direction tangential to the path at P is

$$F = \cos \alpha = mv\frac{dv}{ds} \tag{15.8}$$

Multiplying both sides of Eq. (15.8) by ds and integrating from point A to point B, we get

$$W = \int_{s_A}^{s_B} F \cos \alpha \, ds = \tfrac{1}{2}mv_B^2 - \tfrac{1}{2}mv_A^2 \tag{15.9}$$

Equation (15.9) means that as the particle moves from A to B along *any* path, the work done by the tangential component ($F \cos \alpha$) of the resultant force is equal to the change in kinetic energy of the particle.

Another way of looking at this is that, in using Eq. (15.5), the components of force and displacement must be parallel to each other for work to be done.

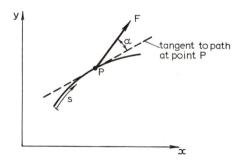

Fig. 15.2 Curvilinear motion of a particle.

For example, if F_x and F_y are the components of F along the x and y axes, respectively, then the equations of motion for the particle shown in Fig. 15.2 are

$$F_x = m\frac{dv_x}{dt} \tag{15.10}$$

$$F_y = m\frac{dv_y}{dt} \tag{15.11}$$

where v_x and v_y are the components of velocity along the x and y axes, respectively. Multiplying Eq. (15.10) by dx, multiplying Eq. (15.11) by dy, integrating, and adding to give the total work done W, we get

$$\begin{aligned} W &= \int_A^B (F_x\,dx + F_y\,dy) \\ &= \tfrac{1}{2}mv_{xB}^2 - \tfrac{1}{2}mv_{xA}^2 + \tfrac{1}{2}mv_{yB}^2 - \tfrac{1}{2}mv_{yA}^2 \\ &= \tfrac{1}{2}mv_B^2 - \tfrac{1}{2}mv_A^2 \end{aligned}$$

or $$W = T_B - T_A \tag{15.12}$$

Again, this equation is identical to Eq. (15.7).

By way of illustration, we can consider the situation in Fig. 15.3, which shows a particle moving from A to B under the action of a force F having constant magnitude and direction. In this case, the work done by the force F is the product of F and the displacement x in the direction of F; i.e., W is equal to Fx. It can also be seen that if we take the component of F in the direction of motion of the particle (i.e., $F\cos\theta$) and multiply this by the displacement s of the particle in the direction of motion, we obtain the same answer; i.e.,

$$W = F\cos\theta(s)$$

or $$W = (F\cos\theta)\left(\frac{x}{\cos\theta}\right) = Fx \tag{15.13}$$

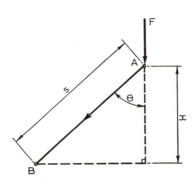

Fig. 15.3 Particle displaced from A to B with force F acting.

Example 15.3 A small sphere of mass m (Fig. 15.4a) is free to roll without resistance around the inside of the horizontal cylindrical duct. If the mass is released from rest at point A, obtain an expression for its velocity when it reaches point B.

Solution. The free-body diagram for the sphere is shown in Fig. 15.4b. The work done by the normal force F_n is zero, while the work done by the force due to gravity mg, as the sphere moves from A to B, is equal to the product of the force and the component of displacement parallel to the line of action of the force. Thus,

$$W = mgr \qquad (a)$$

Since the sphere is released from rest, the velocity at A is zero, and therefore T_A is zero. Hence, substitution of W from Eq. (a) into the work-energy equation (15.7) gives

$$mgr = \tfrac{1}{2}mv_B^2$$

and

$$v_B = \sqrt{2gr}$$

15.3 Conservation of Energy

If a mass is raised above some datum level, it is said to possess *potential energy*. In this case, the potential energy is the work done against the gravita-

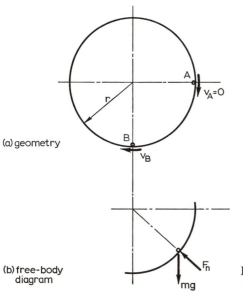

(a) geometry

(b) free-body diagram

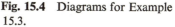

Fig. 15.4 Diagrams for Example 15.3.

tional force acting on the mass. Since gravitational forces always act normal to the surface of the earth, we have a situation similar to that in Fig. 15.3 where to find the work done, we can multiply the force by the displacement in the direction of the force; i.e.,

$$W = -mgh \qquad (15.14)$$

where h is the vertical distance through which the mass has been raised and the negative sign indicates that the displacement was in a direction opposite to the gravitational force mg. Thus, the potential energy V is given by

$$V = -W = mgh \qquad (15.15)$$

Potential energy can also be regarded as the work done in raising a mass and is the amount of work that the mass is able to do in falling back to its orginal height.

Another form of potential energy is the energy stored in a compressed (or extended) spring. If a spring (Fig. 15.5) has a stiffness k (force per unit extension or compression) and is compressed by an amount x from its uncompressed state, then the work done by the force to compress the spring is equal to the potential energy stored in the spring. Then, using Eq. (15.2) and knowing that F is equal to kx, we get

$$V = \int_0^x kx \, dx$$
$$= \tfrac{1}{2}kx^2 \qquad (15.16)$$

This is the amount of potential energy possessed by the compressed spring and is the amount of work that the force exerted by the spring can do as the spring returns to its uncompressed state.

Having introduced the physical significance of the various forms of me-

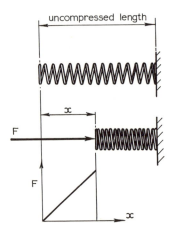

Fig. 15.5 Spring.

chanical energy, we can now introduce the principle of the *conservation of energy*. It has been pointed out that energy exists in many different forms. However, although energy can be converted from one form to another, it cannot be created or destroyed. Thus, we may say that the total energy possessed by a system of moving bodies is at every instant constant (energy is conserved) provided that no energy is rejected to or received from a source external to the system.

Many problems in mechanics may be solved by application of this principle of the conservation of energy. The chief danger lies in the possibility of overlooking the fact that the conditions of the problem may lead to a change of form of some of the energy possessed by the system. For example, when a body slides on a surface, work is done by the frictional force. The energy used is converted into heat and in practice cannot be reconverted into mechanical energy.

On the other hand, the work done in raising a body or compressing a spring provides potential energy which can be recovered, and thus if a problem involves only potential and kinetic energy, we can apply conservation of energy and write

$$T_A + V_A = T_B + V_B \tag{15.17}$$

or
$$-\Delta V = \Delta T \tag{15.18}$$

which means that the loss in potential energy is equal to the gain in kinetic energy or vice versa.

Example 15.4   Apply conservation of energy to Example 15.3.

Solution.   Referring to Fig. 15.4, the loss in potential energy is given by

$$-\Delta V = -(V_B - V_A) = -(0 - mgr) = mgr \tag{a}$$

and the gain in kinetic energy is given by

$$\Delta T = \tfrac{1}{2}mv_B^2 \tag{b}$$

Substitution of Eqs. (a) and (b) into Eq. (15.18) gives

$$mgr = \tfrac{1}{2}mv_B^2$$

or
$$v_B = \sqrt{2gr}$$

15.4   Work Done by Conservative Forces

Gravitational and spring forces are special kinds of forces called *conservative forces*, because the work done by these forces provides potential energy which can be recovered and the mechanical energy is conserved. On the other hand, the work done by a frictional force, for example, is converted into heat and cannot readily be recovered in the form of mechanical energy. A frictional force is known as a nonconservative force. Returning to the problem of a shuffleboard disk (Example 15.1), we saw that the frictional force eventually brought the disk to rest when it initially possessed kinetic energy. After this, the heat generated by the frictional force could not, in practice, be reconverted to mechanical energy in order to return the velocity of the dish to its original value.

In general, we can say that potential energy is the negative of the work done by conservative forces, and the examples of conservative forces of interest in dynamics are spring forces and forces due to gravity. This can be shown mathematically as follows.

It has been seen that the work done by a force F which is acting on a particle as it moves along a curvilinear path from point A to point B is given by

$$W = \int_{s_A}^{s_B} F \cos \alpha \, ds$$
$$= \int_A^B (F_x \, dx + F_y \, dy) \tag{15.19}$$

Depending on the conditions, the value of this integral might be affected by the path followed in going from A to B. For example, if a person pushes a box from A to B around a semicircular path on a horizontal floor, the frictional resistance F_f would be constant and always oppose the motion. The work done by the person would be given by

$$W = F_f \pi r \tag{15.20}$$

where r is the radius of the semicircle. If the person were to have pushed the box direct from A to B, then the work done would be given by

$$W = F_f(2r) \tag{15.21}$$

which is less than that given by Eq. (15.20). Thus, the path taken in this example affects the amount of work done.

Alternatively, returning to the situation shown in Fig. 15.3, where the force F was constant in magnitude and direction, the work done was given by

$$W = Fx \tag{15.22}$$

and the path taken from A to B was irrelevant.

Figure 15.6 shows paths 1 and 2 in the xy plane for a particle moving from A to B (path 1) and returning from B to A (path 2). If this diagram represented a person pushing a box around a floor, then the total work done in going from A to B and returning to A would not be zero. Thus,

$$W = \oint (F_x \, dx + F_y \, dy) \neq 0 \tag{15.23}$$

where the symbol $\oint$ means the line integral around the closed path. The net

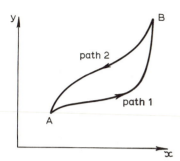

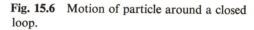

Fig. 15.6 Motion of particle around a closed loop.

work in going around the loop has been converted into heat as a result of friction between the box and the floor. Alternatively, if Fig. 15.6 represented the motion of a particle in a vertical plane, the work done by the force due to gravity in going around the closed path would be zero since the net displacement in the direction of the force would be zero. Thus,

$$W = \oint (F_x \, dx + F_y \, dy) = 0 \qquad (15.24)$$

Under these circumstances, the work done by the force of gravity in going from A to B is recovered in going from B to A, and whatever energy was gained by the system as a result of the work done by the force in going from A to B is lost in going from B to A.

The forces involved if Eq. (15.24) applies are called conservative forces, and mechanical energy has been conserved. The forces involved if Eq. (15.23) applies are nonconservative forces, and the work done has been converted into some nonmechanical form of energy. When the integral of a function around a closed path is zero, the function is an exact differential, and we can write, for conservative forces only,

$$F_x \, dx + F_y \, dy = -dV \qquad (15.25)$$

where V is called the potential energy and the negative sign indicates that potential energy is a measure of the capacity for doing work.

Combining Eqs. (15.19) and (15.25) gives

$$W = - \int_{V_A}^{V_B} dV = V_A - V_B \qquad (15.26)$$

This equation shows that the work done on a particle by a conservative force depends only on the change in potential energy and not on its absolute value.

Substitution of Eq. (15.26) into the work-energy equation (Eq. 15.7) gives

$$T_A + V_A = T_B + V_B \qquad (15.27)$$

which is the same as Eq. (15.17) and means that, for conservative forces, the loss in potential energy is equal to the gain in kinetic energy or vice versa.

Example 15.5   Figure 15.7a shows a vertical pole down which a mass of 1 kg is free to slide. At the bottom of the pole is a compression spring 1 m in length and having a stiffness of 1 kN/m. If the mass is released from rest at the height shown, what is the maximum amount by which the spring is compressed if any energy lost due to initial impact is neglected.

Solution.   Figure 15.7b shows the position of the mass when its velocity has become zero. Since the only two forces acting on the mass, the gravitational force and the spring force, are conservative, conservation of energy applies, and Eq. (15.27) can be written in the form

$$T_A + V_{Ag} + V_{As} = T_B + V_{Bg} + V_{Bs} \tag{a}$$

where V_{Ag} is the potential energy due to the gravitational force at point A and V_{Bs} is the potential energy due to the spring force when the mass is at point B. If V_{Bg} is taken as zero, then V_{Ag} will be $(9 + x)mg$. When the mass is at A, V_{As} is zero, and when the mass reaches B, V_{Bs} is $\frac{1}{2}kx^2$. Since T_A and T_B are both zero, Eq. (a) reduces to

$$(9 + x)mg = \tfrac{1}{2}kx^2 \tag{b}$$

Substitution of $m = 1$ kg and $k = 1000$ N/m gives, after rearrangement,

$$500x^2 - 9.81x - 88.3 = 0$$

Thus,

$$x = \frac{9.81 \pm \sqrt{(9.81)^2 + 4(8.83)(500)}}{1000}$$

$$= 0.43 \text{ m}$$

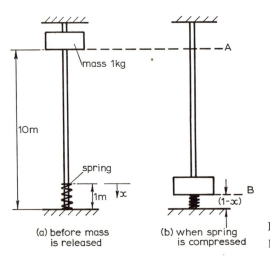

(a) before mass
is released

(b) when spring
is compressed

Fig. 15.7 Diagrams for Example 15.5.

15.5 Work Done by Nonconservative Forces

There are situations when both conservative and nonconservative forces are present in a problem. In this case, it is more convenient to write the work-energy equation (15.12) in the form

$$W_n + W_c = T_B - T_A \qquad (15.28)$$

where W_n is the work done by the nonconservative forces and W_c is the work done by the conservative forces. Since the work done by the conservative forces is the negative of the change in potential energy between A and B, Eq. (15.28) becomes

$$W_n - \Delta V = T_B - T_A$$

or
$$W_n = T_B - T_A + V_B - V_A \qquad (15.29)$$

Example 15.6 A body of mass m is initially at rest on an inclined plane when a constant force F_p is applied as shown in Fig. 15.8a. The coefficient of friction between the body and the inclined plane is μ. Assuming that F_p is of sufficient magnitude to move the body a distance s up the plane, obtain an expression for the final velocity of the body.

Solution. From the free-body diagram in Fig. 15.8b, the frictional force is $\mu(mg \cos \theta + F_p \sin \theta)$, and therefore the work done by this frictional force is given by

$$W_f = -\mu(mg \cos \theta + F_p \sin \theta)s \qquad (a)$$

The component of the force F_p in the direction of motion is $F_p \cos \theta$, and therefore the work done by all the externally applied (nonconservative) forces, including friction, is given by

$$W_n = F_p s \cos \theta - \mu(mg \cos \theta + F_p \sin \theta)s \qquad (b)$$

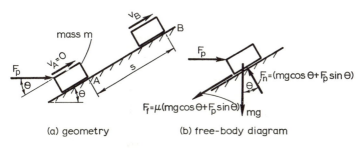

(a) geometry   (b) free-body diagram

Fig. 15.8 Diagrams for Example 15.6.

The final height of the mass above its starting point is $s \sin \theta$, and therefore the change in potential energy as the mass moves from A to B is given by

$$V_B - V_A = mgs \sin \theta \tag{c}$$

Substituting Eqs. (b) and (c) into Eq. (15.29) and setting T_A equal to zero and T_B equal to $\frac{1}{2}mv_B^2$, we get,

$$v_B = \left\{ 2gs \left[\left(\frac{F_p}{mg} - \mu \right) \cos \theta - \left(1 + \frac{\mu F_p}{mg} \right) \sin \theta \right] \right\}^{1/2} \tag{d}$$

Problems

15.1 Solve problem 14.5 using energy methods.

15.2 Solve problem 14.8 using energy methods.

15.3 Solve problem 14.12 using energy methods.

15.4 Find the final velocity of the block in problem 14.16 using energy methods.

15.5 A toy car is projected along a horizontal roadway with an initial velocity v_0. After traveling for 1 m it enters a "loop-the-loop" having a diameter of 0.2 m (Fig. P15.5). If the resistance to motion is constant at 0.1 N and the mass of the car is 0.15 kg, calculate the minimum value of v_0 for the car to reach the top of the loop without leaving the track.

15.6 A vehicle is powered by an engine developing a constant useful power of 100 kW. If its weight is 12 kN and air resistance is negligible, obtain the velocity of the vehicle after traveling 400 m, starting from rest. What would be the maximum velocity of the vehicle if air resistance were $5v^2$, where v is the velocity of the vehicle?

15.7 A capstan, having a drum 0.3 m in diameter and rotating at a constant frequency of $1\ \mathrm{s}^{-1}$, is used to raise an anchor weighing 10 kN. The anchor rope, which is attached to the anchor, first passes over a 150-mm-diameter guide roller and is then wrapped two full turns around the capstan drum (Fig. P15.7). Find the least tension which has to be exerted at the free end of the rope in order to raise the anchor at a constant speed if the coefficient of friction between the rope and the drum is 0.1 and if the frictional torque

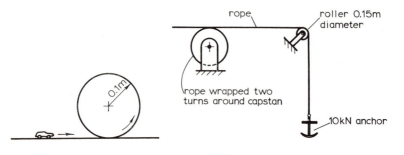

rope   roller 0.15m diameter

rope wrapped two turns around capstan

10kN anchor

0.1m

Fig. P15.5   **Fig. P15.7**

at the guide roller is 15 N·m. Determine the power required to drive the capstan and the rate of doing useful work in raising the anchor.

15.8 The block of mass m shown in Fig. P15.8 is resting on a horizontal surface. The coefficients of static and dynamic friction are μ_s and μ_d, respectively. A horizontal spring of stiffness k (force per unit deflection) is connected to the block, and the free end of the spring is slowly moved away from the block until the block starts to move and is then held stationary. Obtain an expression for the distance d through which the block slides before coming to rest.

15.9 In the amusement park loop ride shown in Fig. P15.9, the car is released from rest at point A, a distance of 15 m above point B. Neglecting friction and wind resistance, determine the velocities of the car when it reaches points B and C.

15.10 In the loop ride shown in Fig. P15.10, the car is initially held against a spring of stiffness k which is compressed an amount δ and then released. Obtain an expression for the velocity of the car when it reaches point B. Neglect friction and wind resistance.

15.11 A model rocket fired vertically achieves a velocity of 75 m/s at burnout, which takes place at a height of 80 m. Neglecting air resistance, determine the maximum altitude the rocket will achieve.

15.12 Figure P15.12 shows a small block of mass m suspended from an unextended spring of stiffness k and an inextensible wire. If the wire is cut, determine the maximum displacement of the mass and the maximum tension in the spring.

15.13 Figure P15.13 shows a simple pendulum which consists of a small sphere of mass m connected to a weightless inextensible string 2 m in length. If the sphere is released from rest in the position shown, determine the velocities of the sphere when it reaches points B and C.

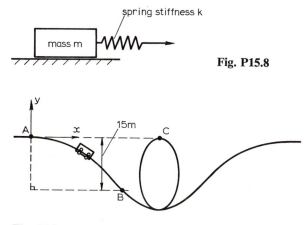

Fig. P15.8

Fig. 15.9

15.14 Figure P15.14 shows two bodies A and B connected by an inextensible rope. The coefficient of friction between body A and the inclined surface is 0.2. If the bodies are released from rest, determine their velocities when the 2-kN body has fallen 5 m. Neglect friction in the pulley.

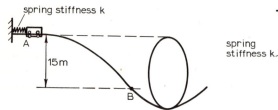

Fig. P15.10

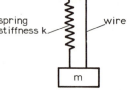

Fig. P15.12

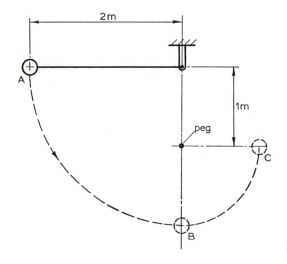

Fig. P15.13

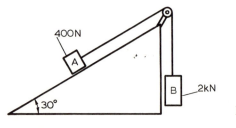

Fig. P15.14

16

Momentum Applied to the Dynamics of Particles

16.1 Introduction

In many problems, particularly those involving impact, the forces acting on particles are a function of time. In these situations, it is necessary to employ the concept of linear momentum. The derivation of impulse and momentum equations, which are the first integral of the equations of motion with respect to time, are introduced in this chapter, which also deals with the law of conservation of linear momentum.

16.2 Conservation of Linear Momentum for a Particle

For a particle of constant mass, Newton's second law of motion may be written in terms of components along the x and y axes, respectively, as follows:

$$F_x = m\ddot{x} = \frac{d(m\dot{x})}{dt} = \frac{d(mv_x)}{dt} \tag{16.1}$$

$$F_y = m\ddot{y} = \frac{d(m\dot{y})}{dt} = \frac{d(mv_y)}{dt} \tag{16.2}$$

The product of mass and velocity is defined as the *linear momentum*, and thus Eqs. (16.1) and (16.2) mean that the components of force along the x and y directions, respectively, are equal to the rate of change of linear momentum in the x and y directions, respectively. Since velocity is a vector quantity and mass is a scalar quantity, linear momentum is a vector quantity.

If the component of force acting on the particle in the x direction is zero, then integration of Eq. (16.1) gives

$$m\dot{x} = \text{constant} \qquad (16.3)$$

or $$m\dot{x}_A = m\dot{x}_B \qquad (16.4)$$

where $\dot{x}_A$ and $\dot{x}_B$ are the velocities of the particle at points A and B, respectively. These equations mean that if the force acting on a particle in the x direction is zero, then the linear momentum of the particle is conserved in the x direction. Similarly, if the resultant force in the y direction is zero, then the linear momentum is conserved in the y direction.

16.3   Conservation of Momentum for a Group of Particles

Figure 16.1 shows one of a group of particles; it has an external force F_e acting upon it and another force F_i which represents the resultant of the forces due to interactions (such as elastic or rigid connections and mutual forces of attraction) between this particle and the remaining particles. At a given instant, the particle has coordinates (x, y) with respect to some inertial reference system. The equation of motion for the x direction is

$$F_{ex} + F_{ix} = m\ddot{x} \qquad (16.5)$$

where m is the mass of the particle. If similar equations are written for all the particles, then summing these equations gives

$$\sum F_{ex} + \sum F_{ix} = \sum (m\ddot{x}) \qquad (16.6)$$

Referring to the forces of interaction F_i between particles, according to Newton's third law, a force exerted by one particle on another will have an equal and opposite force exerted on the first particle by the second. This means

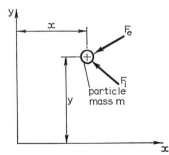

Fig. 16.1   Forces acting on one particle in a group.

that in a group of particles the summation of all such forces will be zero, and thus the term $\sum F_{ix}$ in Eq. (16.6) will be zero. Hence, Eq. (16.6) becomes

$$\sum F_{ex} = \sum (m\ddot{x}) \qquad (16.7)$$

Now if x_c is the x coordinate of the center of mass of the system of particles, then

$$\sum (mx) = (\sum m)x_c \qquad (16.8)$$

Differentiating Eq. (16.8) twice with respect to time, we get

$$\sum (m\ddot{x}) = (\sum m)\ddot{x}_c \qquad (16.9)$$

where $\ddot{x}_c$ is the acceleration of the center of mass of the system and $\sum m$ is its total mass. Finally, eliminating $\sum(m\ddot{x})$ from Eqs. (16.7) and (16.9) gives

$$\sum F_{ex} = (\sum m)\ddot{x}_c \qquad (16.10)$$

A similar equation of motion exists for the y direction. Thus,

$$\sum F_{ey} = (\sum m)\ddot{y}_c \qquad (16.11)$$

Equations (16.10) and (16.11) mean that the acceleration of the center of mass of a system of particles multiplied by the total mass is equal to the sum of the external forces acting on the particles.

For convenience, Eqs. (16.10) and (16.11) are usually written in the same form as the equations of motion for a single particle; that is,

$$F_x = m\ddot{x}_c \qquad (16.12)$$

$$F_y = m\ddot{y}_c \qquad (16.13)$$

where m is understood to mean the total mass of the group of particles and F_x and F_y are understood to mean the components of the resultant of all the external forces acting on the system of particles.

If the resultant force in the x direction acting on the system of particles is zero, then from Eq. (16.7) we have, by integration,

$$\sum (m\dot{x}) = \text{constant} \qquad (16.14)$$

or from Eq. (16.10), by integration,

$$(\sum m)\dot{x}_c = \text{constant} \qquad (16.15)$$

These last two equations express, for the x direction, the law of conservation of linear momentum for a group of particles. Obviously, if the resultant force in the y direction is also zero, then linear momentum is also conserved in the y direction. Thus, if there are no external forces acting on a system of particles, then the center of mass of the system must move at constant speed or remain stationary.

Example 16.1   A girl scout whose mass is 50 kg is at one end of a canoe of mass 100 kg (Fig. 16.2a). The distance from the pier to the girl scout is 5 m. If the canoe is not tied to the pier and the water is still, how far is she from the pier after walking the length of the canoe? Neglect any friction between the canoe and the water.

Solution.   Before the girl scout starts walking, the center of mass of the system is stationary; i.e., $\dot{x}_c$ is zero. Since no external forces are acting in the x direction, from Eq. (16.15),

$$(\sum m)\dot{x}_c = \text{constant} = 0 \qquad (a)$$

where the constant is zero since $\dot{x}_c$ is initially zero. Integrating Eq. (a) gives

$$x_c = \text{constant} \qquad (b)$$

which means that the center of mass of the system must remain in the same position throughout. To find the position of the center of mass we use

$$(\sum m)x_c = \sum (mx) \qquad (c)$$

Thus, taking moments about the pier end of the canoe, we get

$$(100 + 50)x_c = 100(2.5) + 50(5)$$

$$x_c = \frac{250 + 250}{150} = 3.33 \text{ m}$$

Now, since the center of mass of the girl and canoe cannot move, it must remain 3.33 m from the pier after she has walked to the other end of the canoe.

Figure 16.2b shows this situation, and it can be seen that the final distance x between the girl scout and the pier will be given by

$$x = 3.33 - 1.67 = 1.67 \text{ m}$$

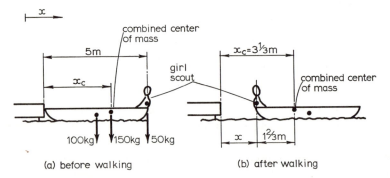

(a) before walking   (b) after walking

Fig. 16.2   Geometry for Example 16.1.

because when the girl is at one end of the canoe, the distance between her and the combined center of mass is 1.67 m.

16.4   Impulse and Impact

If Eq. (16.1) is integrated with respect to time, we get

$$\int_{t_1}^{t_2} F_x \, dt = \int_{mv_{x1}}^{mv_{x2}} d(mv) = mv_{x2} - mv_{x1} \tag{16.16}$$

The left-hand side of this equation is the *time integral* of the force component and is therefore equal to the change in *linear momentum* of the particle.

The *angular momentum of a particle about a fixed point is defined as the moment of its linear momentum.* Thus, if a particle has a linear momentum mv (Fig. 16.3), then the angular momentum p_o about O is

$$p_o = m(v_y x - v_x y) \tag{16.17}$$

Differentiating Eq. (16.17) with respect to time gives

$$\dot{p}_o = m(\dot{v}_y x - v_y \dot{x} - \dot{v}_x y - v_x \dot{y}) \tag{16.18}$$

or since

$$\dot{v}_y = \ddot{y}, \quad \dot{v}_x = \ddot{x}, \quad \dot{x} = v_x, \quad \dot{y} = v_y$$

then

$$\dot{p}_o = m(\ddot{y}x - \ddot{x}y) = F_y x - F_x y \tag{16.19}$$

Since $F_y x - F_x y$ is the moment M_o of the external forces about O, Eq. (16.19) becomes

$$\dot{p}_o = M_o \tag{16.20}$$

Then the *rate of change of angular momentum of a particle about a fixed point is equal to the moment about the same point of the external forces applied to the*

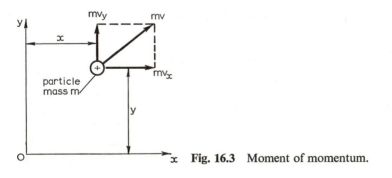

Fig. 16.3   Moment of momentum.

particle. Integrating Eq. (16.20) gives

$$\int_{t_1}^{t_2} M_o \, dt = \int_{p_{o1}}^{p_{o2}} dp_o = p_{o2} - p_{o1} \qquad (16.21)$$

Thus, the *time integral of M_o is equal to the change in angular momentum about O.*

A finite change of momentum may be produced, for example, by a small force or moment acting for an appreciable interval of time or by a very large force or moment acting for a short interval of time. When the interval of time is very short, the time integral of the force is known as the *impulse,* and the force is known as an *impulsive force.* Impulsive forces occur in collisions, in explosions, in the striking of a nail by a hammer, or in the action of a pile driver. If an impulse is denoted by a caret ($^\wedge$) above the impulsive force, then from Eq. (16.16), as $\Delta t \to 0$,

$$\hat{F}_x = \lim_{\Delta t \to 0} \int F_x \, dt = mv'_x - mv_x \qquad (16.22)$$

where v_x and v'_x are the velocities immediately before and immediately after impact, respectively.

Similarly, in the y direction,

$$\hat{F}_y = \lim_{\Delta t \to 0} \int F_y \, dt = mv'_y - mv_y \qquad (16.23)$$

Thus, the *impulse of a force acting on a particle is equal to the change in linear momentum of the particle.* It should be noted that although Δt approaches zero, it is assumed that the force is sufficiently large so that the time integral of the force exists and is finite. Another way of saying this is that although Δt approaches zero, so that no change in position of the particle occurs, the force is large enough so that an instantaneous change in velocity occurs.

From Eq. (16.21), as $\Delta t \to 0$,

$$\hat{M}_o = \lim_{\Delta t \to 0} \int M_o \, dt = p'_o - p_o \qquad (16.24)$$

and the *moment of an impulse about O is equal to the change in the moment of linear momentum (angular momentum) about O.*

In situations which utilize Eqs. (16.22)–(16.24) it is assumed that the impulsive forces and moments are much larger than any finite force in the system (such as gravity) so that the latter may be neglected.

As an example, we shall consider the collision or impact between two spheres moving with different velocities along the same straight line (Fig. 16.4). The behavior of the colliding bodies during the complete period of impact will depend on the properties of the materials of which they are made. Figures 16.4a and 16.4c show the two spheres immediately before impact and immediately after impact. Before and after contact, each sphere moves independently of the other. During contact (Fig. 16.4b) a force of interaction

acts along the common normal between the two bodies for a short interval of
time. A time plot of the magnitude of the contact force (Fig. 16.5) consists of
an approach period, of duration Δt_a, during which the bodies deform locally
in the vicinity of the contact point, followed by a separation period, of dura-
tion Δt_s. In the ideal case of elastic impact, $\Delta t_a = \Delta t_s$, while for an inelastic
impact, $\Delta t_a > \Delta t_s$. For the so-called case of plastic impact, the bodies remain
in contact with no elastic recovery after the approach or compression period,
and Δt_s is zero.

The impulsive forces acting on each of the two masses must be equal in
magnitude but opposite in direction. Thus, from the free-body diagrams
shown in Fig. 16.6, the impact equations along the direction of impact are

$$-\hat{F} = m_1(v'_1 - v_1) \tag{16.25}$$

$$\hat{F} = m_2(v'_2 - v_2) \tag{16.26}$$

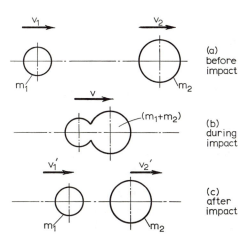

(a)
before
impact

(b)
during
impact

(c)
after
impact

Fig. 16.4 Collision between two
spheres.

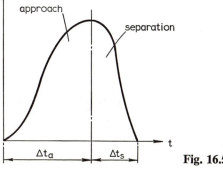

Fig. 16.5 Impulsive force.

m_1' m_2 **Fig. 16.6** Free-body diagrams during impact.

Eliminating $\hat{F}$ between Eqs. (16.25) and (16.26) gives

$$m_1v_1 + m_2v_2 = m_1v_1' + m_2v_2' \tag{16.27}$$

If the two spheres were treated as a group of particles, then the impulsive forces would become equal and opposite internal forces, so that the external force acting on the system of particles would be zero. Thus, the linear momentum of the system would be conserved, and application of Eq. (16.14) would also give Eq. (16.27).

In general, because of energy losses that occur due to the impact, the impulsive force $\int F_s\, dt$ during the period of separation differs from the impulsive force $\int F_a\, dt$ during the approach period. This, in turn, causes the relative velocities after separation to differ from those before separation. The ratio of the magnitude of these impulsive forces or the ratio of the postcollision and precollision relative velocities, along the common normal at the point of contact, is called the coefficient of restitution and is denoted by e. Thus,

$$e = \frac{\int F_s\, dt}{\int F_a\, dt} = \frac{v_2' - v_1'}{v_1 - v_2} \tag{16.28}$$

or $e = \dfrac{\text{relative velocity of separation}}{\text{relative velocity of approach}}$

The value of e has been found to depend on the geometry of the bodies, the centrality of impact, the relative approach velocity, and the materials. In the ideally elastic case, where $\hat{F}_s = \hat{F}_a$, $e = 1$. For plastic impact, $e = 0$, while in the inelastic case, $0 < e < 1$.

For the special case when v_2 is zero, Eqs. (16.27) and (16.28) can be solved to give

$$v_1' = \frac{v_1(m_1 + m_2 e)}{m_1 + m_2} \tag{16.29}$$

$$v_2' = \frac{v_1(1 + e)m_1}{m_1 + m_2} \tag{16.30}$$

In this case, before impact, the kinetic energy of the system is $\frac{1}{2}m_1v_1^2$. After impact, the total kinetic energy of the system becomes $\frac{1}{2}m_1v_1'^2 + \frac{1}{2}m_2v_2'^2$. Thus, the change in kinetic energy ΔT is given by

$$\Delta T = \frac{1}{2}m_1v_1^2 - \frac{1}{2}(m_1v_1'^2 + m_2v_2'^2) \tag{16.31}$$

Substitution of Eqs. (16.29) and (16.30) into Eq. (16.31) gives

$$\Delta T = \frac{(1 - e^2)m_1 v_1^2}{2[1 + (m_1/m_2)]} \qquad (16.32)$$

For given values of m_1 and v_1, the loss of kinetic energy will depend on the mass ratio m_1/m_2 and on the value of e. The smaller the mass ratio, the greater will be the loss of kinetic energy. The smaller the value of e, the greater the loss in kinetic energy. Thus, the energy loss is maximized when m_2 is very large compared with m_1 and when a plastic impact ($e = 0$) occurs.

It follows from the work-energy equation that the loss of kinetic energy due to impact is equal to the work done in deforming the two bodies. If the two bodies are perfectly plastic, they will be permanently deformed and the work done converted into heat energy. There will be no potential energy stored in the material due to elastic deformation, and there will be no tendency for either body to regain its original shape. Hence, the two bodies will remain in contact and move on together with reduced kinetic energy. The impact between two lead spheres or two clay spheres approximates to plastic impact.

If the colliding bodies are perfectly elastic, the whole of the work done in deforming the bodies will be stored in the bodies as potential energy, and there will be no conversion of kinetic energy to heat energy. Immediately after the instant at which the two bodies are moving with the same velocity, the bodies will begin to regain their original shapes, the potential energy being reconverted into kinetic energy and the two bodies ultimately separating. In this case, the impulses on each of the colliding bodies will have the same magnitude while the bodies are separating as they had during the first stage. Hence, the change in momentum of each body during the second phase will be equal to the change in momentum during the first phase.

Example 16.2 A sphere of mass 30 kg moving at 3 m/s overtakes and collides with another sphere of mass 20 kg moving at 1 m/s in the same direction. Find the velocities of the two masses after impact and the loss of kinetic energy during impact (a) when the impact is plastic, (b) when it is elastic, and (c) when $e = 0.6$.

Solution. (a) Plastic impact. The two masses move together after impact and move with a common velocity, $v_1' = v_2'$. From Eq. (16.27), which expresses the law of conservation of linear momentum, we get

$$v_1' = \frac{m_1 v_1 + m_2 v_2}{m_1 + m_2}$$

$$v_1' = \frac{30(3) + 20(1)}{30 + 20} = \frac{90 + 20}{50} = \frac{110}{50} = 2.2 \text{ m/s}$$

The total kinetic energy before impact is given by

$$T = \tfrac{1}{2}(m_1 v_1^2 + m_2 v_2^2)$$
$$= \tfrac{1}{2}[30(3)^2 + 20(1)^2] = 145 \text{ N·m}$$

The total kinetic energy after impact is given by

$$T' = \frac{v'^2}{2}(m_1 + m_2)$$

$$= \frac{(2.2)^2}{2}(30 + 20) = 121 \text{ N·m}$$

Thus, the loss of kinetic energy is

$$\Delta T = T - T' = 145 - 121 = 24 \text{ N·m}$$

(b) Elastic impact. From the conservation of linear momentum, Eq. (16.27), we get

$$m_1 v_1 + m_2 v_2 = m_1 v_1' + m_2 v_2'$$

or $\qquad\qquad 30(3) + 20(1) = 30 v_1' + 20 v_2'$

and $\qquad\qquad\qquad 110 = 30 v_1' + 20 v_2' \qquad\qquad$ (a)

For an elastic impact, $e = 1$, and from Eq. (16.28), we get

$$e = \frac{v_2' - v_1'}{v_1 - v_2}$$

or $\qquad\qquad\qquad 1 = \frac{v_2' - v_1'}{3 - 1}$

and $\qquad\qquad\qquad v_2' - v_1' = 2 \qquad\qquad$ (b)

Eliminating v_1' between Eqs. (a) and (b) gives

$$110 = 30(v_2' - 2) + 20 v_2'$$

or $\qquad\qquad\qquad v_2' = 3.4 \text{ m/s} \qquad\qquad$ (c)

Hence, from Eq. (b),

$$v_1' = 3.4 - 2 = 1.4 \text{ m/s} \qquad\qquad \text{(d)}$$

The initial kinetic energy T is 145 N·m, and T' is given by

$$T' = \tfrac{1}{2}(m_1 v_1'^2 + m_2 v_2'^2)$$
$$= \tfrac{1}{2}[30(1.4)^2 + 20(3.4)^2]$$
$$= 145 \text{ N·m}$$

Thus, ΔT is zero, as it should be for an elastic impact.

(c) $e = 0.6$. From Eq. (16.28), we get

$$e = \frac{v_2' - v_1'}{v_1 - v_2}$$

and therefore,

$$v_2' - v_1' = e(v_1 - v_2) = 0.6(2) = 1.2 \text{ m/s} \qquad (e)$$

From the conservation of linear momentum, Eq. (16.27),

$$m_1 v_1 + m_2 v_2 = m_1 v_1' + m_2 v_2'$$

or $30(3) + 20(1) = 30v_1' + 20v_2'$

and $110 = 30v_1' + 20v_2'$

Substitution for v_1' from Eq. (e) gives

$$110 = 30(v_2' - 1.2) + 20v_2'$$

or $v_2' = 2.92 \text{ m/s}$

and $v_1' = 2.92 - 1.2 = 1.72 \text{ m/s}$

As before, the kinetic energy T before impact is 145 N·m, and T' is given by

$$\begin{aligned} T' &= \tfrac{1}{2}(m_1 v_1'^2 + m_2 v_2'^2) \\ &= \tfrac{1}{2}[30(1.72)^2 + 20(2.92)^2] \\ &= 129.64 \text{ N·m} \end{aligned}$$

Thus, the loss in kinetic energy is given by

$$\Delta T = T - T' = 145 - 129.64 = 15.36 \text{ N·m}$$

Example 16.3   A drop forging operation is performed by releasing a hammer of mass m_1 from a height h above a metal workpiece which rests on an anvil of mass m_2 (Fig. 16.7). The impact of the hammer against the workpiece is used to deform the workpiece by converting some of the kinetic energy of the falling hammer into energy that is dissipated in changing the shape of the worhpiece. Determine the values of e and m_2 such that a maximum portion of the kinetic energy will be used in deforming the workpiece.

Solution.   To maximize the transfer of energy from the hammer to the work-piece, the separation force after impact should be as small as possible. This occurs if $e = 0$, which implies that the hammer does not rebound. Thus, from Eq. (16.32),

$$\Delta T = \frac{m_1 v_1^2}{2[1 + (m_1/m_2)]}$$

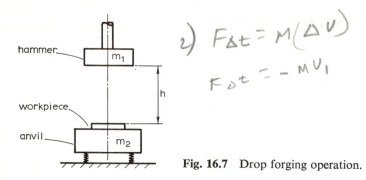

$$2)\ F\Delta t = M(\Delta v)$$

$$F_{\partial t} = -Mv_1$$

Fig. 16.7 Drop forging operation.

Hence, to maximize ΔT, the anvil should have as large a mass m_2 as possible, and the maximum value of ΔT is $\frac{1}{2}m_1 v_1^2$.

Problems

16.1   Two particles A and B, each of mass m, are attached to the ends of a light stiff spring AB, and the system is placed on a smooth horizontal table. A blow of impulse $\hat{F}$ is applied to A in the direction AB. Show that the maximum compression of the spring is $\hat{F}/\sqrt{2mk}$, where k is the force required to compress the spring by unit length.

16.2   A ship weighing 300 MN traveling at 1 m/s is to be brought to a stop by a small tugboat applying a force of 100 kN. If the resistance of the water to the motion of the ship is neglected, how long will it take to stop the ship? Use the principles of impulse and momentum.

16.3   A man weighing 700 N jumps into a rowboat floating stationary by the beach. If the man has an initial horizontal velocity of 3 m/s and the boat weighs 1.4 kN, what is the velocity of the boat and man together just after he has landed in the boat?

16.4   A pile of weight 10 kN is driven into the ground by repeatedly dropping a weight of 1.5 kN on it through a height of 1 m. Each blow causes a movement of the pile of 100 mm. Assuming no rebound occurs during the impact, find the mean resistance of the ground to the penetration of the pile. What is the loss of energy at each impact?

16.5   A bullet of mass 0.05 kg is fired horizontally into a block of wood of mass 20 kg. The block is resting on a horizontal surface, and the coefficient of friction between the block and the surface is 0.2. How far will the block slide if the velocity of the bullet is 500 m/s?

16.6   A pool player projects the cue ball at 0.8 m/s to strike the eight ball head on. If the two balls have the same mass and the impact is elastic, what is the velocity of the eight ball after impact? (Note: Treat the balls as particles, and neglect friction.)

$$3)\ M_1 v_1 + M_2 v_2 = (M_1 + M_2) v'$$

$$6)\ M_1 v_1 + M_2 \dots$$

$$M = A$$

16.7 A horizontal table of mass 20 kg is supported by springs of combined stiffness 1 kN/m. A mass of 2 kg is dropped onto the table from a height of 2 m. If the impact is elastic, find the maximum compression of the springs.

16.8 An 0.06-kg bullet is fired into a 500-N wooden block which is suspended as a simple pendulum by a 5-m cord (Fig. P16.8). If, after striking the block, the pendulum swings through an angle of 10°, determine the initial velocity of the bullet. What proportion of the initial energy was lost during the impact?

16.9 Two particles, each of mass 2 kg, are connected by a light rigid bar 1 m in length. If the arrangement is dropped from rest in the position shown in Fig. P16.9 and mass A impacts with the rigid steel plate and then rebounds ($e = 0.6$), determine the angular velocity of the bar (a) when the maximum deformation of A has occurred and (b) immediately after impact.

16.10 An artillery gun of mass m_1 is mounted on a frictionless horizontal track. The gun, whose barrel makes an angle of θ with the track, fires a shell of mass m_2. If the muzzle velocity is v_m (i.e., the velocity of the shell relative to the gun), obtain an expression for the recoil velocity of the gun and the absolute velocity of the shell. Also determine the initial angle that the trajectory of the shell makes with the horizontal.

16.11 A particle of mass m_1 lies midway along a tube of length $2b$ and mass m_2. The tube, which is closed at both ends, lies on a smooth horizontal table. The coefficient of restitution between the particle and the tube is e. If m_1 is given an initial velocity v_0 along the tube, obtain expressions for (a) the velocities of the particle and tube after the first impact, (b) the energy loss during the first impact, and (c) the time required for the particle to arrive back at its initial position and traveling in its original direction.

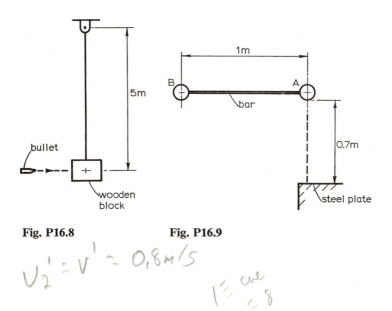

Fig. P16.8   Fig. P16.9

17

Dynamics of Rigid Bodies

17.1 Introduction

In the previous three chapters, the motion of particles due to the application of external forces was studied. There, each body was assumed to be either infinitesimally small so that its mass and all the external forces acting were concentrated at the same point, or the orientation of the body in space was unimportant so that it was not necessary to take into account the body's shape or the exact location of the forces. In this chapter, we shall be concerned with a study of the relationships between the forces that act on any rigid body and the corresponding translational and rotational motions of the body.

Only two equations of motion are needed to determine either the planar motion of a single particle or the planar motion of the center of mass of a group of particles. For the planar motion of a rigid body, an additional equation of motion is needed in order to determine the rotational motion of the body.

17.2 Planar Motion of a Rigid Body

Figure 17.1a shows the free-body diagram for a rigid body moving in the xy plane. Since a rigid body is merely a collection of particles with no relative linear motion between them, the motion of the center of mass of the body can be determined from the translational equations of motion in the x and y

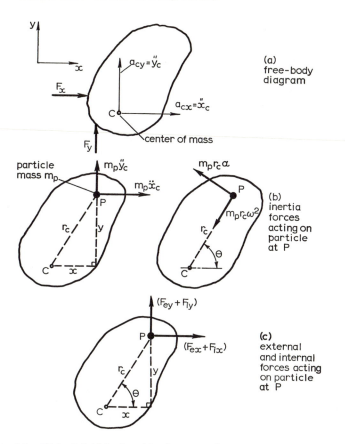

Fig. 17.1 Rigid body with planar motion.

directions as follows:

$$F_x = ma_{cx} = m\ddot{x}_c \tag{17.1}$$

$$F_y = ma_{cy} = m\ddot{y}_c \tag{17.2}$$

where m is the total mass of the body; F_x and F_y are the sums of the components of the external forces in the x and y direction, respectively; and a_{cx} and a_{cy} are the components of the acceleration of the center of mass in these two directions, respectively.

To develop the rotational equation of motion, we can examine first the motion of a single particle of mass m_p located at P (Fig. 17.1b). This motion consists of the rotation of point P about the center of mass C superimposed on the motion of the center of mass. Thus, the total inertia forces for the particle will be the sum of the individual inertia force components shown in the two diagrams in Fig. 17.1b. Now if F_{ex} and F_{ix} represent the x components óf the external and internal forces acting on the particle, respectively, and F_{ey}, F_{iy} represent the y components, the free-body diagram for the particle will be as shown in Fig. 17.1c. Taking moments about C and equating the moments of the forces in Fig. 17.1c to the sum of the moments of all the individual inertia forces in Fig. 17.1b, we get

$$x(F_{ey} + F_{iy}) - y(F_{ex} + F_{ix}) = m_p r_c^2 \alpha + x m_p \ddot{y}_c - y m_p \ddot{x}_c \qquad (17.3)$$

Summing for all particles and remembering that all internal forces act in equal and opposite pairs and therefore cancel, we get

$$\sum (x F_{ey} + y F_{ex}) = \sum m_p r_c^2 \alpha + (\sum x m_p) \ddot{y}_c - (\sum y m_p) \ddot{x}_c \qquad (17.4)$$

The left-hand side of Eq. (17.4) is the sum of the moments M_c of the external forces about the center of mass, and, by definition of the center of mass, $\sum x m_p$ and $\sum y m_p$ are both zero. Thus,

$$M_c = (\sum m_p r_c^2)\alpha \qquad (17.5)$$

Finally, the term $\sum m_p r_c^2$ is defined as the moment of inertia I_c about an axis through the center of mass of the body and perpendicular to the xy plane. The moment of inertia can also be regarded as the second moment of mass and compared with the second moment of area introduced earlier in the work on hydrostatics.

If we consider a body to have a continuous distribution of mass instead of a discrete distribution of mass particles, then the total mass may be divided into small elements of mass Δm, and as Δm approaches zero, the moment of inertia I_c is given by

$$I_c = \int r_c^2 \, dm \qquad (17.6)$$

Hence, Eq. (17.5) becomes

$$M_c = I_c \alpha \qquad (17.7)$$

This is the rotational equation of motion for the body and, together with Eqs. (17.1) and (17.2), can be used to analyze the planar motion of a rigid body. However, before presenting some examples, it is necessary to explain how the moment of inertia of a rigid body is obtained.

17.3 Moment of Inertia

For simple-shaped bodies, such as slender rods, right circular cylinders, and spheres, the moment of inertia of the body can be evaluated by direct integration using Eq. (17.6). For more complicatedly shaped bodies, it is necessary to divide the body into a number of simple-shaped elements and sum the individual moments of inertia about a common axis.

Example 17.1 Determine the moment of inertia of a long slender rod of mass m and length l about an axis through the center of mass and perpendicular to the length of the rod (Fig. 17.2).

Solution. If ρ is the density of the material and A the cross-sectional area of the rod, then for an element of length dx the mass of the element dm is equal to $\rho A \, dx$, and

$$I_c = \int_{-l/2}^{l/2} x^2 \rho A \, dx = \left(\rho A \frac{x^3}{3} \right)_{-1/2}^{1/2} = \frac{\rho A l^3}{12}$$

Since the total mass m of the bar is equal to $\rho A l$,

$$I_c = \frac{ml^2}{12}$$

Example 17.2 Determine the moment of inertia of a homogeneous right circular cylinder of mass m, length l, and radius r about its longitudinal axis z (Fig. 17.3).

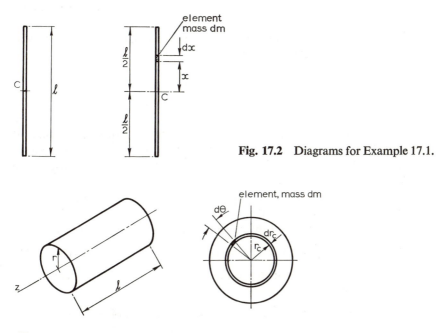

Fig. 17.2 Diagrams for Example 17.1.

Fig. 17.3 Diagrams for Example 17.2.

Solution. If ρ is the density of the material, then in cylindrical coordinates the mass dm of a small element is given by

$$dm = \rho \, dV = \rho l r_c \, d\theta \, dr_c$$

and
$$I_z = \int r_c^2 \, dm = \rho l \int_0^{2\pi} \int_0^r r_c^3 \, dr_c \, d\theta = \frac{\rho l \pi r^4}{2}$$

Since the mass of the cylinder is $\rho l \pi r^2$,

$$I_z = \frac{mr^2}{2}$$

17.3.1   Radius of Gyration

Frequently, the moment of inertia of a body about a given axis is presented in the form

$$I = mk^2 \tag{17.8}$$

where m is the mass of the body and k is a constant for the body. The constant k has the unit of length and is called the *radius of gyration* about the axis under consideration. If a single particle of mass m_p were located a distance k from an axis, the moment of inertia of the particle about that axis would be $m_p k^2$.

It is sometimes convenient to remember that for a plane surface k^2 is equal to the second moment of area divided by the area.

17.3.2   Parallel-Axis Theorem

If the moment of inertia of a body about an axis passing through its center of mass (centroidal axis) is known, then its moment of inertia about another axis, parallel to the centroidal axis, can readily be determined by use of the *parallel-axis theorem*. We shall consider a body of mass m having an x, y coordinate system whose origin is at an arbitrary point O and an x', y' coordinate system whose origin is at the center of mass C of the body (Fig. 17.4). The moment of inertia about the axis passing through O is given by

$$I_o = \int r_o^2 \, dm = \int (x^2 + y^2) \, dm = \int [(x_c + x')^2 + (y_c + y')^2] \, dm \tag{17.9}$$

where (x_c, y_c) are the coordinates of the center of mass in the x, y coordinate system. Expanding Eq. (17.9) gives

$$I_o = (x_c^2 + y_c^2) \int dm + \int (x'^2 + y'^2) \, dm + 2x_c \int x' \, dm + 2y_c \int y' \, dm$$

$$I_o = d^2 \int dm + \int r_o'^2 \, dm + 2x_c \int x' \, dm + 2y_c \int y' \, dm \tag{17.10}$$

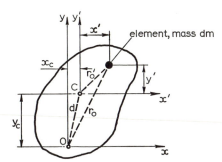

Fig. 17.4 Geometry for parallel-axis theorem.

where d is the distance between the axes passing through O and C. The second term on the right-hand side of Eq. (17.10) represents the moment of inertia I_c of the body about an axis through the center of mass, while the last two terms are zero by definition of the center of mass. Thus,

$$I_o = I_c + md^2 \tag{17.11}$$

Equation (17.11) expresses the parallel-axis theorem.

Example 17.3   Determine the moment of inertia of a slender rod of mass m and length l about an axis perpendicular to its length and through one of its ends (Fig. 17.5).

Solution.   From Example 17.1, the moment of inertia of the rod about an axis through its center of mass was found to be given by

$$I_c = \frac{ml^2}{12}$$

Thus, from Eq. (17.11) we get

$$I_o = I_c + md^2 = \frac{ml^2}{12} + m\left(\frac{l}{2}\right)^2$$

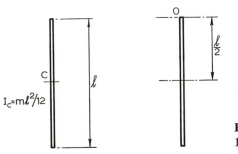

Fig. 17.5 Diagrams for Example 17.3.

or
$$I_o = \frac{ml^2}{3}$$

Example 17.4 A steel crank consists of a disk of thickness 10 mm and radius 100 mm and a cylindrical pin of radius 25 mm and length 20 mm, as shown in Fig. 17.6. Determine the moment of inertia about the axis through O perpendicular to the plane of the disk. The density of steel is 7850 kg/m³.

Solution. The moment of inertia of the crank is the sum of the moments of inertia of the disk and pin about the axis passing through O. For the disk, from Example 17.2,

$$I_{do} = \frac{mr^2}{2} = \tfrac{1}{2}\rho(\pi r^2 l)r^2 = \tfrac{1}{2}(7850)\pi(0.01)(0.1)^4$$

$$= 0.012 \text{ kg·m}^2$$

while for the pin

$$I_{po} = \frac{mr^2}{2} + md^2 = m\left(\frac{r^2}{2} + d^2\right)$$

$$= \rho\pi r^2 l\left(\frac{r^2}{2} + d^2\right) = 7850\pi(0.025)^2(0.02)\left[\frac{(0.025)^2}{2} + (0.075)^2\right]$$

$$= 0.002 \text{ kg·m}^2$$

Thus,

$$I_o = I_{do} + I_{po} = 0.014 \text{ kg·m}^2$$

Example 17.5 If the second moment of area of a circular section of radius r about a diameter is $\pi r^4/4$, find the radius of gyration of a thin circular disk of radius r about a diameter. Hence, find the moment of inertia of a cylinder of radius r and length l about a transverse axis passing through its center of mass.

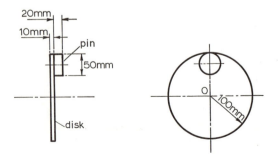

Fig. 17.6 Steel crank.

Solution.   The required radius of gyration is given by

$$k^2 = \frac{\text{second moment of area}}{\text{area}}$$

$$= \frac{\pi r^4/4}{\pi r^2} = \frac{r^2}{4} \tag{a}$$

Thus, for the elemental disk of thickness dx and mass dm shown in Fig. 17.7, the moment of inertia dI_y about the y axis is

$$dI_y = \frac{r^2}{4} \, dm + x^2 \, dm \tag{b}$$

Now if ρ is the density of the material,

$$dm = \rho \pi r^2 \, dx \tag{c}$$

and substitution of Eq. (c) into Eq. (b) and integration give

$$I_y = \rho \pi r^2 \left(\int_{-l/2}^{+l/2} \frac{r^2}{4} dx + \int_{-l/2}^{+l/2} x^2 \, dx \right)$$

$$= \rho \pi r^2 \left[\frac{r^2}{4}(l) + \tfrac{1}{3}\left(\frac{l^3}{8} + \frac{l^3}{8}\right) \right]$$

$$= \rho \pi r l^2 \left(\frac{r^2}{4} + \frac{l^2}{12}\right)$$

Since the mass m of the cylinder is given by $\rho \pi r^2 l$,

$$I_y = m\left(\frac{r^2}{4} + \frac{l^2}{12}\right)$$

17.4 Translational Motion of a Rigid Body

If a rigid body moves in such a way that any line in the body remains parallel to its original orientation, then the angular velocity of the body is zero, and

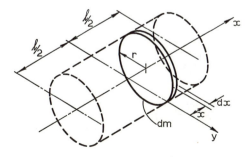

Fig. 17.7 Diagram for Example 17.5.

the body has only translational motion. In this case, the rotational equation of motion about the center of mass reduces to

$$M_c = 0 \tag{17.12}$$

where M_c is the sum of the moments of the external forces about the center of mass. The translational equations of motion, however, remain the same as before. That is,

$$F_x = m\ddot{x}_c \tag{17.13}$$

$$F_y = m\ddot{y}_c \tag{17.14}$$

where F_x, F_y are the sums of the components of the external forces and $\ddot{x}_c$, $\ddot{y}_c$ are the components of acceleration of the center of mass.

If the resultant external force acting on a translating rigid body is denoted by F_r, then

$$F_r = \sqrt{(F_x)^2 + (F_y)^2} \tag{17.15}$$

and from Eq. (17.12), the moment of F_r about the center of mass must be zero. Thus, the line of action of the resultant force must pass through the center of mass.

Example 17.6   A body of mass m_1 rests on a wedge of mass m_2 as shown in Fig. 17.8a. The coefficient of static friction between the body and the wedge is μ. What is the maximum force F_p that can be applied to the wedge so that the body does not move relative to the wedge?

Solution.   Free-body and inertia-force diagrams for the body alone and for the body and wedge together are shown in Figs. 17.8b and 17.8c, respectively. If a is the acceleration of the wedge, then resolving parallel to the wedge surface, for the body alone, we get

$$\mu F_{n1} + F_{w1} \sin 30° = m_1 a \cos 30° \tag{a}$$

and resolving perpendicular to the wedge surface, we get

$$F_{n1} - F_{w1} \cos 30° = m_1 a \sin 30° \tag{b}$$

From Eq. (b) and since $F_{w1} = m_1 g$,

$$\mu F_{n1} = F_{w1}\left(\frac{\sqrt{3}}{2} + \frac{a}{2g}\right) \tag{c}$$

and from Eq. (a),

$$\mu F_{n1} = F_{w1}\left(\frac{\sqrt{3}a}{2g} - \frac{1}{2}\right) \tag{d}$$

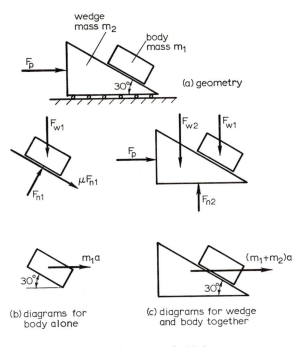

Fig. 17.8 Diagrams for Example 17.6.

Dividing Eq. (c) by Eq. (d) and rearranging, we get

$$a = g\left(\frac{\mu\sqrt{3} + 1}{\sqrt{3} - \mu}\right) \tag{e}$$

Resolving horizontally for the body and wedge together gives

$$F_p = (m_1 + m_2)a$$

or

$$F_p = (m_1 + m_2)g\left(\frac{\mu\sqrt{3} + 1}{\sqrt{3} - \mu}\right) \tag{f}$$

Example 17.7 A locomotive side connecting rod has the dimensions shown in Fig. 17.9a. Assuming that the rod is a straight uniform bar whose mass is 225 kg and that it is supported by frictionless pins, determine the reaction forces at the pins when the locomotive is running at 25 m/s.

Solution. Since the rod has translational motion only, its angular velocity is zero, and the equations of motion are

$$F_x = m\ddot{x}_c, \quad F_y = m\ddot{y}_c, \quad M_c = 0$$

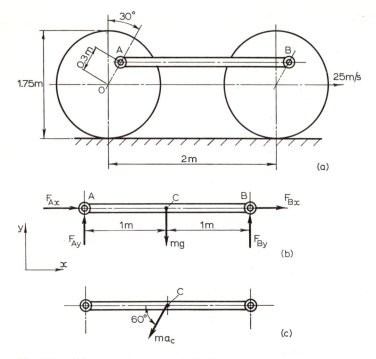

Fig. 17.9 Diagrams for Example 17.7.

The acceleration of the center of mass of the rod is equal to the acceleration of points A and B on the wheels. Since the locomotive speed is constant, the angular velocity of the wheel is constant, and the acceleration of point A has a radial component only, directed from A to O and of magnitude $0.3\omega^2$. The angular velocity ω of the wheel is

$$\omega = \frac{v}{r} = \frac{25}{1.75/2} = 28.57 \text{ rad/s} \qquad \text{(a)}$$

Thus, the acceleration a_c of the center of mass of the rod is given by

$$a_c = 0.3(28.57)^2 = 244.9 \text{ m/s}^2 \qquad \text{(b)}$$

and results in the inertia force ma_c directed as shown in Fig. 17.7c. Hence, the equations of motion for the rod (see Figs. 17.7b and 17.7c) are as follows: For the x direction,

$$F_{Ax} + F_{Bx} = m\ddot{x}_c = -225(244.9 \cos 60°) = -27{,}551 \qquad \text{(c)}$$

and for the y direction,

$$F_{Ay} + F_{By} - mg = m\ddot{y}_c = -225(244.9 \sin 60°) = -47{,}720 \qquad \text{(d)}$$

Taking moments about C, we get

$$F_{Ay} = F_{By} \tag{e}$$

Solving Eqs. (d) and (e) gives

$$F_{Ay} = F_{By} = -22,756 \text{ N} \tag{f}$$

Since all the independent equations from rigid body dynamics have been used, the horizontal components F_{Ax} and F_{Bx} of the reactions are indeterminate.

17.5   Rotation of a Rigid Body About a Fixed Axis

Figure 17.10 shows a rigid body rotating about a line perpendicular to the xy plane passing through a point O which is not necessarily the location of the center of mass of the body. The radial and tangential components of the inertia force acting on a particle of mass m_p, located at point P are shown in the figure. Remembering that the internal forces due to interactions between the particles which make up the rigid body all cancel, taking moments about O and summing for all particles, we get

$$M_o = (\sum m_p r_o^2)\alpha \tag{17.16}$$

where M_o is the moment of the external forces acting on the body about point O. The term $\sum m_p r_o^2$ is defined as the moment of inertia I_o of the body about point O. Thus,

$$M_o = I_o\alpha \tag{17.17}$$

This equation is similar to Eq. (17.7) except that in this case moments are taken about the fixed axis of rotation at O rather than the location of the center of mass of the body.

In addition to Eq. (17.17), the translational equations of motion, Eqs. (17.1) and (17.2), still apply.

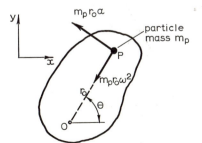

Fig. 17.10   Rigid body rotating about O.

Example 17.8 A slender uniform rod is standing vertically on a rough horizontal surface. The rod is given a small angular displacement, and the angle through which it will rotate before its lower end slips on the surface is 45°. Find the coefficient of static friction between the rod and the surface.

Solution. Figure 17.11a shows the external forces acting on the rod after it has rotated through an angle θ from the vertical. Figure 17.11b shows, for the same instant, the inertia forces acting due to the radial and tangential accelerations of the center of mass, and Fig. 17.11c shows the inertia moment about O, the lower end of the rod. Using Eq. (17.17), or equating the moments about O, for Figs. 17.11a and 17.11c, we get

$$I_o\ddot{\theta} = mga \sin \theta$$

or
$$\ddot{\theta} = \frac{mga}{I_o} \sin \theta \tag{a}$$

Resolving vertically for Figs. 17.11a and 17.11b, we get, for the motion of the center of mass,

$$F_n = mg - ma\dot{\theta}^2 \cos \theta - ma\ddot{\theta} \sin \theta \tag{b}$$

Resolving horizontally, we get

$$F_f = ma\ddot{\theta} \cos \theta - ma\dot{\theta}^2 \sin \theta \tag{c}$$

Before we can proceed to find the values of F_f and F_n in terms of θ, we need an expression for $\dot{\theta}$. This expression may be obtained as follows. Since

$$\ddot{\theta} = \frac{d\dot{\theta}}{dt} = \frac{d\dot{\theta}}{d\theta}\frac{d\theta}{dt} = \dot{\theta}\frac{d\dot{\theta}}{d\theta} \tag{d}$$

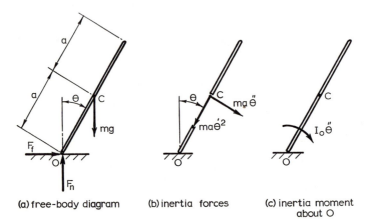

(a) free-body diagram   (b) inertia forces   (c) inertia moment about O

Fig. 17.11 Diagrams for Example 17.8.

substitution of Eq. (d) into Eq. (a) and rearrangement give

$$\int_0^{\dot\theta} \dot\theta \, d\dot\theta = \frac{mga}{I_o} \int_0^\theta \sin\theta \, d\theta$$

or

$$\left(\frac{\dot\theta^2}{2}\right)_0^{\dot\theta} = \frac{mga}{I_o}(-\cos\theta)_0^\theta$$

Thus,

$$\dot\theta^2 = \frac{2mga}{I_o}(1 - \cos\theta) \tag{e}$$

Substitution for $\ddot\theta$ from Eq. (a) and for $\dot\theta$ from Eq. (e) into Eqs. (b) and (c) and writing $I_o = mk_o^2 = \frac{4}{3}ma^2$, we get

$$F_n = mg[1 - \tfrac{3}{4}(1 + 2\cos\theta - 3\cos^2\theta)] \tag{f}$$

$$F_f = \tfrac{3}{4}mg(3\sin\theta\cos\theta - 2\sin\theta) \tag{g}$$

The critical condition when the rod slips is given by

$$F_f = \mu F_n \tag{h}$$

and substitution of Eqs. (f) and (g) into Eq. (h) gives

$$\mu = \frac{3\sin\theta\cos\theta - 2\sin\theta}{\tfrac{1}{4} - \tfrac{3}{2}\cos\theta + \tfrac{9}{4}\cos^2\theta}$$

or, when $\theta = 45°$,

$$\mu = 0.27$$

Example 17.9   Figure 17.12a shows a uniform rod of length $2a$ and mass m rotating with a constant angular velocity ω about a fixed horizontal axis O passing through the center of mass of the rod. Derive expressions for the

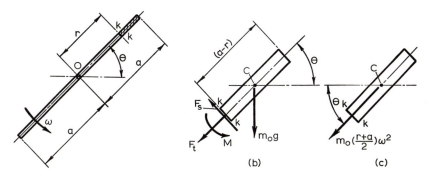

Fig. 17.12   Diagrams for Example 17.9.

maximum tension force, the maximum shear force, and the maximum bending moment in the rod at section k-k, a distance r from the axis of rotation.

Solution. The free-body and inertia-force diagrams for the portion of the
rod beyond section k-k are shown in Figs. 17.12b and 17.12c, respectively.
Denoting the tension force by F_t, the shear force by F_s, and the bending
moment by M and resolving forces parallel to the rod, we get

$$F_t + m_o g \sin \theta = m_o \left(\frac{r + a}{2} \right) \omega^2 \tag{a}$$

where m_o is the mass of the portion of the rod beyond section k-k. Since

$$m_o = \frac{m(a - r)}{2a} \tag{b}$$

then

$$F_t = \frac{m(a - r)}{2a} \left[\frac{(a + r)\omega^2}{2} - g \sin \theta \right] \tag{c}$$

The maximum value of F_t will occur when $\sin \theta = -1$, i.e., when $\theta = 270°$.
Thus,

$$F_{t\,max} = \frac{m(a - r)}{2a} \left[\frac{(a + r)}{2} \omega^2 + g \right] \tag{d}$$

Resolving forces perpendicular to the rod, we get

$$F_s - m_o g \cos \theta = 0 \tag{e}$$

and substitution of Eq. (b) into Eq. (e) gives

$$F_s = \frac{m}{2a}(a - r)g \cos \theta \tag{f}$$

The maximum values of F_s will occur when $\cos \theta = \pm 1$, i.e., when $\theta = 0$ or
180°. Thus,

$$F_{s\,max} = \pm \frac{mg}{2a}(a - r) \tag{g}$$

Finally, taking moments about the center of mass C of the portion of the
rod shown beyond section k-k gives

$$M - F_s \left(\frac{a - r}{2} \right) = 0 \tag{h}$$

and substitution of Eq. (f) into Eq. (h) gives

$$M = \frac{m(a - r)^2}{4a} g \cos \theta \tag{i}$$

Again, the maximum values of M will be given when $\cos \theta = \pm 1$, i.e., when $\theta = 0$ or $180°$. Thus,

$$M_{\max} = \pm \frac{mg}{4a}(a - r)^2 \tag{j}$$

17.6 General Planar Motion of a Rigid Body

The equations of motion for the general case of planar motion of a rigid body have already been derived in Section 17.2. We have now given examples for bodies subjected to translational motion only and for bodies subjected to rotational motion about a fixed point. In this section, we shall present examples of the general case, where the body is both translating and rotating. However, a word of explanation regarding the use of the moment equation [Eq. (17.7)] is required. In our work, a technique has been introduced whereby two diagrams, one for the external forces and one for the inertia forces, are drawn and then the corresponding force components equated. This technique can be used in deriving equations of motion for general planar motion of a rigid body as follows.

Figure 17.13 shows force diagrams for a body which has an angular acceleration $\ddot{\theta}$ and where the absolute acceleration of an arbitrarily chosen point P in the body is aligned with the x axis. Figure 17.13a shows the external forces acting on the body, while Fig. 17.13b shows the inertia forces due to the components of acceleration of the center of mass, together with the inertia moment $I_c\ddot{\theta}$ due to the angular acceleration of the body about its center of mass.

The equations for the translational motion of the center of mass are obtained as usual. Thus, resolving radially, we get

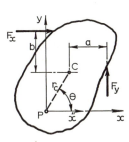

(a) free-body diagram

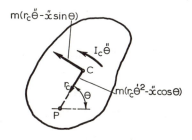

(b) inertia forces and moment for center of mass

Fig. 17.13 Planar motion of a rigid body.

$$F_x \cos \theta + F_y \sin \theta = -m(r_c \dot\theta^2 - \ddot x \cos \theta) \tag{17.18}$$

and resolving tangentially, we get

$$F_y \cos \theta - F_x \sin \theta = m(r_c \ddot\theta - \ddot x \sin \theta) \tag{17.19}$$

However, since the inertia force diagram was drawn to include the inertia moment about the center of mass (Fig. 17.13b), a moment equation can be obtained by taking moments about any convenient point, in much the same way as we did for the statics of rigid bodies. To illustrate this, we can first take moments about the center of mass C. Thus,

$$F_y a - F_x b = I_c \ddot\theta \tag{17.20}$$

Equation (17.20) corresponds to Eq. (17.7), derived earlier. Alternatively, taking moments about an arbitrary point P, we get

$$F_y(a + r_c \cos \theta) - F_x(b + r_c \sin \theta) = I_c \ddot\theta + m(r_c^2 \ddot\theta - \ddot x r_c \sin \theta)$$

$$F_y a - F_x b + r_c(F_y \cos \theta - F_x \sin \theta) = I_c \ddot\theta + m r_c^2 \ddot\theta - m \ddot x r_c \sin \theta \tag{17.21}$$

Substitution of Eq. (17.19) into Eq. (17.21) gives Eq. (17.20). Hence, moments can be taken about any point when the inertia-force diagram includes the inertia moment about the center of mass.

In the special case where point P coincides with a fixed axis of rotation O, then $\ddot x$ is zero, and Eq. (17.21) reduces to

$$M_o = (I_c + m r_c^2)\ddot\theta = I_o \ddot\theta \tag{17.22}$$

which corresponds to Eq. (17.17).

Example 17.10   A cable is wrapped around the inner drum of a wheel and pulled horizontally with a force of 200 N as shown in Fig. 17.14a. The wheel is resting on a horizontal surface and has a mass of 50 kg and a moment of inertia, about its axis, of 0.4 kg·m². If the static coefficient of friction μ_s between the wheel and the surface is 0.2 and the dynamic coefficient of friction μ_d is 0.15, determine the linear and angular accelerations of the wheel.

Solution.   We shall first assume that rolling occurs without slipping. The free-body and inertia-force diagrams are shown in Fig. 17.14. Taking moments about the center of mass C gives

$$I_c \alpha = F_f r_0 - F_p r_i \tag{a}$$

Resolving horizontally, we get

$$ma = F_p - F_f \tag{b}$$

Resolving vertically, we get

$$F_n - mg = 0 \tag{c}$$

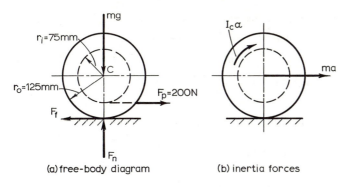

(a) free-body diagram   (b) inertia forces

Fig. 17.14   Diagrams for Example 17.10.

Eliminating F_f from Eqs. (a) and (b) and writing $a = r_0\alpha$ for no slipping, we get

$$I_c\alpha = r_0(F_p - mr_0\alpha) - F_p r_i$$

or

$$\alpha = \frac{F_p(r_0 - r_i)}{I_c + mr_o^2}$$

$$= \frac{200(125 - 75) \times 10^{-3}}{0.4 + 50(125 \times 10^{-3})^2}$$

$$= 8.47 \text{ rad/s}^2 \qquad\qquad (d)$$

From Eq. (b), we get

$$F_f = F_p - mr_0\alpha$$

$$= 200 - 50(125 \times 10^{-3})(8.47)$$

$$= 147 \text{ N} \qquad\qquad (e)$$

and from Eq. (c),

$$F_n = mg$$

$$= 50(9.81)$$

$$= 490.5 \text{ N}$$

The maximum friction force that the contact between the wheel and the surface can withstand before slipping occurs is given by

$$F_{f \text{ max}} = \mu_s F_n$$

$$= 0.2(490.5)$$

$$= 98.1 \text{ N} \qquad\qquad (f)$$

Since the value of F_f obtained above is greater than $F_{f\text{max}}$, the wheel will slip.

When the wheel rolls and slips, the free-body and inertia-force diagrams and the resulting equations of motion do not change, but we can now say that

$$F_f = \mu_d F_n$$
$$= 0.15(490.5)$$
$$= 73.6 \text{ N} \tag{g}$$

Substitution of this value into Eq. (a) gives

$$\alpha = \frac{F_f r_0 - F_p r_i}{I_c}$$
$$= \frac{73.6(125) - 200(75)}{0.4 \times 10^3}$$
$$= -14.5 \text{ rad/s}^2$$

The negative sign indicates that the angular acceleration is counterclockwise. Finally, from Eq. (b) we get

$$a = \frac{F_p - F_f}{m}$$
$$= \frac{200 - 73.6}{50}$$
$$= 2.53 \text{ m/s}^2$$

Example 17.11   For the simple-engine mechanism in the position shown in Fig. 17.15a, determine the forces exerted on the piston pin A and the crank pin B for a constant angular velocity of the crank of 377 rad/s. The connecting rod AB has a mass of 7 kg with its center of mass C in the position shown and a moment of inertia I_c of 0.2 kg·m². Neglect friction and the pressure on the 3-kg piston. Assume that forces due to gravity are negligible compared to the forces due to acceleration, and use the approximate equation for piston acceleration.

Solution.   The free-body and inertia-force diagrams for the piston and connecting rod are shown in Figs. 17.15b and 17.15c, respectively. Resolving horizontally for the piston (Fig. 17.15b) gives

$$F_{Ax} = m_1 a_{Ax} \tag{a}$$

From the analysis of a simple-engine mechanism given earlier,

$$a_{Ax} \simeq \omega^2 r \left(\cos \theta + \frac{r}{l} \cos 2\theta \right) \tag{b}$$

where r is the radius of the crank, l is the length of the connecting rod, ω is the

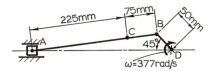

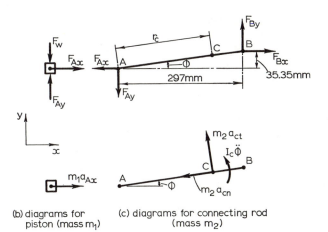

(a) simple-engine mechanism

(b) diagrams for piston (mass m_1)   (c) diagrams for connecting rod (mass m_2)

Fig. 17.15 Diagrams for Example 17.11.

angular velocity of the crank, and θ is the angle of crank rotation past top dead center. Substitution of the appropriate values for the present example gives

$$a_{Ax} = \frac{(377)^2(50)(0.707)}{1000}$$

$$= 5.02 \text{ km/s}^2$$

From Eq. (a),

$$F_{Ax} = 3(5.02 \times 10^3) = 15.1 \text{ kN}$$

Again, from the earlier analysis of a simple-engine mechanism,

$$\dot{\phi} = \frac{r}{l}\omega \cos \theta = \frac{50}{300}(377)(0.707) = 44.42 \text{ rad/s}$$

and

$$\ddot{\phi} = -\frac{r}{l}\omega^2 \sin \theta = -\frac{50}{300}(377)^2(0.707) = -16.75 \text{ krad/s}^2$$

and by geometry,

$$\sin \phi = \frac{r}{l} \sin \theta = \frac{50}{300}(0.707) = 0.12$$

and $$\cos \phi = 0.99$$

Substituting these values into expressions for the tangential and normal inertia forces acting at the center of mass (Fig. 17.15c), we get

$$m_2 a_{ct} = m_2(r_c \ddot{\phi} - a_{Ax} \sin \phi)$$

$$= 7\left[-\frac{225}{1000}(16.75) - (5.03 \times 10^3)(0.12)\right]$$

$$= -4.252 \text{ kN}$$

and $$m_2 a_{cn} = m_2(r_c \dot{\phi}^2 - a_{Ax} \cos \phi)$$

$$= 7\left[\frac{225}{1000}(44.42)^2 - (5.03 \times 10^3)(0.99)\right]$$

$$= -31.75 \text{ kN}$$

and for the inertia moment about the center of mass,

$$I_c \ddot{\phi} = 0.2(-16.75)$$

$$= -3.35 \text{ N} \cdot \text{m}$$

Taking moments about B, we get

$$F_{Ay}(297 \times 10^{-3}) - F_{Ax}(35.35 \times 10^{-3}) = I_c \ddot{\phi} - m_2 a_{ct}(75 \times 10^{-3})$$

$$F_{Ay}(297 \times 10^{-3}) = (15.1 \times 10^3)(35.35 \times 10^{-3}) - 3.35$$

$$+ (4.252 \times 10^3)(75 \times 10^{-3})$$

or $$F_{Ay} = 2.86 \text{ kN}$$

Therefore,

$$F_A = \sqrt{F_{Ax}^2 + F_{Ay}^2}$$

$$= \sqrt{(15.1)^2 + (2.16)^2} = 15.37 \text{ kN}$$

Resolving vertically for F_{By}, we get

$$F_{By} - F_{Ay} = m_2 a_{ct} \cos \phi - m_2 a_{cn} \sin \phi$$

or $$F_{By} = 2.86 \times 10^3 - (4.25 \times 10^3)(0.99) + (31.75 \times 10^3)(0.12)$$

$$= 2.46 \text{ kN}$$

Resolving horizontally for F_{Bx}, we get

$$F_{Bx} - F_{Ax} = -m_2 a_{ct} \sin \phi - m_2 a_{cn} \cos \phi$$

or $$F_{Bx} = 15.1 \times 10^3 + (4.25 \times 10^3)(0.12) + (31.75 \times 10^3)(0.99)$$

$$= 47.04 \text{ kN}$$

Finally,

$$F_B = \sqrt{F_{Bx}^2 + F_{By}^2}$$
$$= \sqrt{(47.04)^2 + (2.46)^2}$$
$$= 47.1 \text{ kN}$$

Problems

17.1 For the rectangular prism shown in Fig. P17.1, obtain expressions for the moment of inertia about the centroidal axes x, y, and z.

17.2 For the rectangular prism shown in Fig. P17.1, obtain an expression for the moment of inertia about one of the long edges.

17.3 For the thin rectangular plate shown in Fig. P17.3, find the radii of gyration about the centroidal axes x and y.

17.4 If the plate shown in Fig. P17.3 has an 0.2-m-diameter hole in its center, what would be the radii of gyration about axes x and y?

17.5 Obtain an expression for the moment of inertia of a thin hemispherical shell of radius r (a) about a diameter of its base and (b) about its axis of symmetry.

17.6 Show that the moment of inertia of a uniform solid right circular cone of mass m, height h, and semivertical angle $45°$ about a line through the vertex perpendicular to the axis of symmetry is $\frac{3}{4}mh^2$. Deduce the moment of inertia of the cone about a diameter of its base.

17.7 A motorcyclist is to round a curve of constant radius at a constant speed of 7 m/s. If the coefficient of friction between the tires and the pavement is 0.7, determine the minimum radius of the curve.

17.8 Figure P17.8 shows a 400-kN monorail car which decelerates uniformly from 30 m/s to 10 m/s in 300 m. Determine the normal reaction between the rail and the front wheel when the velocity is 20 m/s.

17.9 If for the monorail car shown in Fig. P17.8 the coefficient of friction between the rail and the wheels is 0.3, determine the maximum acceleration of the

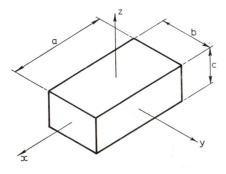

Fig. P17.1

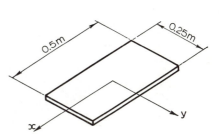

Fig. P17.3

monorail car so that no slipping occurs. Assume (a) rear-wheel drive and (b) front-wheel drive.

17.10 A wheel and axle arrangement is shown in Fig. P17.10 and is supported in frictionless bearings. The mass of the disk-shaped wheel is 8 kg, and its diameter is 0.8 m. The mass of the axle is 4 kg, and its diameter is 0.2 m. A light cord secured to, and wrapped around, the axle supports a mass of 2 kg. Determine the time for the mass to fall a distance of 1 m starting from rest with the cord taut.

17.11 A disk flywheel is mounted on an axle, and the assembly is suspended with its axis horizontal by means of two light strings which are wrapped around the axle as shown in Fig. P17.11. The radius of the axle is r_1 and the radius of the flywheel r_2. The moment of inertia of the axle may be neglected. Initially, the whole system is at rest with the strings vertical and supporting the weight of the wheel and axle. The flywheel is suddenly released, and and simultaneously the upper ends of the strings are moved vertically upward together so that their ends remain on a horizontal line. Show that for the flywheel axis to remain stationary after the flywheel has been released, the upper ends of the strings must be given vertical accelerations upward of $2g(r_1/r_2)^2$.

17.12 The cone described in problem 17.6 can turn freely about a horizontal hinge

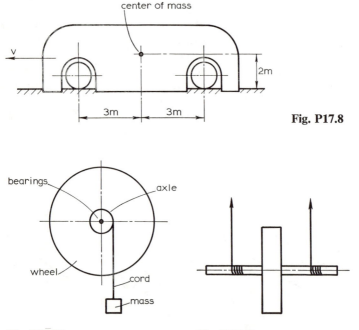

center of mass

2m

3m   3m

Fig. P17.8

bearings

axle

wheel

cord

mass

Fig. P17.10   **Fig. P17.11**

attached to a diameter of its base and is released from rest with its axis horizontal. Show that when the axis makes an angle θ with the downward vertical, the cone exerts on the hinge a force of magnitude $\frac{3}{4}mg(1 + 3\cos^2\theta)^{1/2}$.

17.13 A tumbling barrel hopper feeder is shown in Fig. P17.13 and consists of a 10-N thin cylindrical shell of length 0.35 m and radius 125 mm, supported on four 40-mm-diameter rollers, two at A and two at B. If a 15-N·m torque is applied to each of the two rollers at B, determine the angular acceleration of the hopper.

17.14 Figure P17.14 shows a rigid body which is free to swing about a fixed axis at O. Obtain expressions for the vertical and horizontal components of the reaction at the support if the body is released from rest with an initial value of $\theta = \theta_o$. Neglect any friction at the support.

17.15 A bifilar pendulum consists of a rigid body suspended by two massless vertical strings of length l attached to opposite ends of the body (Fig. P17.15). The rigid body consists of a uniform slender bar of length b and mass m. If the bar is rotated through a small angle about a vertical axis passing through its center of mass and then released, obtain an expression for the angular velocity of motion of the bar.

17.16 Figure P17.16 shows a hub with two blades attached rotating with an angular

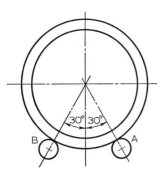

Fig. P17.13

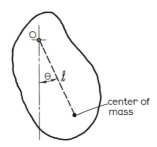

Fig. P17.14

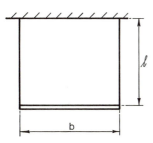

Fig. P17.15

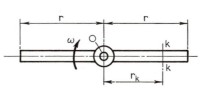

Fig. P17.16

velocity ω about a horizontal shaft at O. For the position shown, obtain expressions for the tension force F_t, the shear force F_s, and the bending moment M in the blade at a section $k\text{-}k$, located a radial distance r_k from O. Assume that the mass per unit length of the blades is m_l.

17.17   The flywheel shown in Fig. P17.17 consists of a thin ring of radius r and mass m connected to a small hub by lightweight spokes. If the flywheel is rotating at a constant angular velocity ω, obtain an expression for the tension force per unit cross-sectional area of the ring.

17.18   Determine the maximum acceleration of the truck shown in Fig. P17.18 so that the crate will not tip.

17.19   Obtain an expression for the maximum velocity v which the monorail car shown in Fig. P17.8 can reach at a distance s, starting from rest, if the rear driving wheels do not slip. The coefficient of friction between the wheels and rail is μ.

17.20   A homogeneous cylinder of mass m and radius r rolls without slipping and starting from rest down an inclined plane (Fig. P17.20). Obtain an expression for the angular velocity of the cylinder in terms of the number of complete revolutions n of the cylinder and the angle of inclination θ of the plane.

17.21   A cylinder of length 100 mm and radius 25 mm has an axle passing through its center of mass perpendicular to its axis of symmetry. The cylinder is placed between parallel inclined rails so that the axle rests across the rails. If the diameter of the axle is 1 mm, the inclination of the rails is 5°, and the cylinder starts from rest, how long will it take the cylinder to roll a distance of 0.5 m? Neglect the mass of the protruding ends of the axle.

17.22   In the hoist shown in Fig. P17.22, the mass of the hoisting drum, motor drive, and beam together is 240 kg with a combined center of mass 1.1 m to the left of the weld attachment at A. If the drum is lowering the 500-kg load with a velocity v which increases uniformly from 0.6 m/s to 6 m/s in 4 s, determine (a) the force in the cable and (b) the bending moment in the beam at A during this interval. The moment of inertia of the drum about its axis is 20 kg·m².

17.23   A uniform circular disk of radius r is projected in its own plane, which is vertical, along a rough horizontal plane. The initial velocity of the center

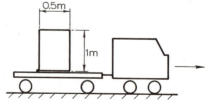

Fig. P17.17   **Fig. P17.18**

of the disk is v, and the initial angular velocity of the disk is ω in the direction which will make the disk return to its starting point. Prove that the disk will stop slipping when it returns to its initial position if $r\omega = 5v$.

17.24   A uniform circular disk is projected, with its plane vertical, along a line of greatest slope of a plane of inclination α to the horizontal. Initially, the disk has no angular velocity, and its center is moving up the plane with a velocity v. If the coefficient of friction between the disk and the plane is $\frac{1}{2}\tan\alpha$, show that slipping ceases after a time $2v/(5g\sin\alpha)$.

17.25   A uniform circular disk of mass m and radius r is at rest, with its plane vertical, on a rough horizontal table, the coefficient of friction at the point of contact being μ. A constant horizontal force F_p is applied to the disk in a line through its center and in its plane. Prove that slipping will occur if $F_p > 3\mu mg$. If $F_p = 6\mu mg$ and the force is applied for a time t and then removed, prove that the disk will continue to slip for a further time t, and find its velocity when slipping ceases.

17.26   The door of an antique car stands open and perpendicular to the length of the car. The car starts off with an acceleration a, and at the same time the door is given an angular velocity ω in the direction toward the front of the car, so as to shut the door. Show that if the door can be regarded as a smoothly jointed uniform rectangular plate of width $2b$, then ω must be at least of magnitude $(3a/2b)^{1/2}$ in order to close the door.

17.27   A uniform plank of mass m_1 is placed symmetrically upon two equal rough uniform cylindrical rollers, each of mass m_2, on a rough plane inclined at an angle α to the horizontal. Assuming no slipping and that the cylinders do not touch, find the initial values of the frictional forces on each cylinder and the acceleration of the plank.

17.28   A uniform rod AB, of mass m and length $6l$, is held horizontally and in contact with a fixed rough horizontal rail C, where $AC = 4l$. The rod is released from rest. At time t after release and before slipping takes place AB makes an angle θ with the horizontal. Show that $2l\dot{\theta}^2 = g\sin\theta$. Show also that the rod begins to slip when θ attains the value $\arctan(\mu/2)$, where μ is the coefficient of friction between the rod and the rail.

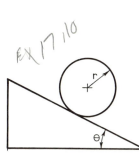

EX 17.10

Fig. P17.20

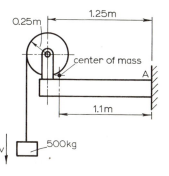

0.25m

1.25m

center of mass

A

1.1m

500kg

Fig. P17.22

17.29   The rectangular block shown in Fig. P17.29 is resting on the edge of a step with its center of mass directly above the edge of the step. If the coefficient of static friction between the block and the step is 0.2 and the block is given a small displacement so that it starts to fall over the step, determine the angle through which it will rotate before it starts to slide relative to the step.

17.30   The uniform disk of radius r and mass m shown in Fig. P17.30 is wound with a flexible light cord which is secured at A. If the coefficient of sliding friction between the disk and the surface is μ, obtain an expression for the force F which will give the center of the disk an acceleration down the plane equal to the acceleration due to gravity, g.

17.31   A slender rod AB of length $2l$ and mass m is freely pivoted about a horizontal axis perpendicular to the rod a distance $l/2$ from A. If the rod is released from a horizontal position, develop an expression for the horizontal reaction at the pivot in terms of m, g, and θ, the angle through which the rod has rotated.

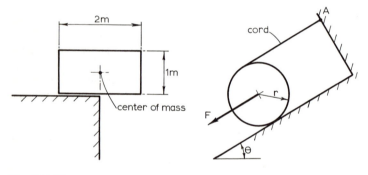

Fig. P17.29   **Fig. P17.30**

18

Work, Power, and Energy Applied to the Dynamics of Rigid Bodies

18.1 Introduction

In Chapter 15 it was pointed out that when the applied forces are either a constant or a function of position, the use of the work-energy equation is often a more convenient alternative for the determination of the velocity of a particle. In this chapter, the work-energy equation is extended to cover the motion of rigid bodies. In our earlier work it was found that a system of external forces acting on a body can always be reduced to a single force whose line of action passes through the center of mass of the body and a moment about the center of mass. Thus, when considering the equations of motion for a rigid body the problem can be separated into (1) a resultant force acting through the center of mass and the resulting motion of the center of mass and (2) a moment about the center of mass and the resulting angular motion of the body about its center of mass. The same procedure can be applied to the work-energy relations.

18.2 Kinetic Energy Due to Translation

For the motion of the center of mass of a body, the body can be considered as a single particle, concentrated at the center of mass, and we can use the equations developed earlier for a particle. Thus, from Eqs. (15.2) and (15.9), the work done W_t due to translation of the center of mass by a force F whose line of action passes through the center of mass is given by

$$W_t = \int F \, ds = \tfrac{1}{2}mv_{cB}^2 - \tfrac{1}{2}mv_{cA}^2 \tag{18.1}$$

where s is the displacement of the center of mass in the direction of the force F, m is the mass, and v_{cA}, v_{cB} are the velocities of the center of mass when it is positioned at points A and B, respectively. It will be noted that we have omitted any work done by internal forces due to interaction between the particles making up the rigid body. Internal forces arise in equal and opposite pairs, and thus the net work done by these forces would be zero since the displacement of all particles would be identical during pure translation of the body. The kinetic energy T_t, due to translational motion of the center of mass, is given by

$$T_t = \tfrac{1}{2}mv_c^2 \tag{18.2}$$

Combining Eqs. (18.1) and (18.2), we get

$$W_t = T_{tB} - T_{tA} = \Delta T_t \tag{18.3}$$

which means that the work done W_t by an external force causing translational motion is equal to the change in kinetic energy of translational motion.

18.3   Kinetic Energy Due to Rotation

For rotation of a rigid body about its center of mass, we again neglect the work done by the internal forces due to interaction between particles. Figure 18.1 shows two particles A and B, in a body which is rotating about O, with the equal and opposite internal forces indicated by F_i. If the body is rotated through an angle $\Delta\theta$ about O, the displacement of line AB can be regarded as a linear displacement together with a rotational displacement of A relative to B. The net work done by the equal and opposite forces F_i due to the linear displacement will be zero. In addition, the displacement of A relative to B due to the rotational component of displacement is at right angles to the direc-

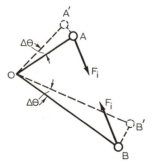

Fig. 18.1   Particles A and B in a rigid body rotating about O.

tion of the internal force acting on A. Hence, the work done by the internal force is again zero. Thus, the net work done by the internal forces due to rotational motion of a rigid body is zero.

The work done W_r by a moment M causing rotation of a rigid body is given by

$$W_r = \int M \, d\theta \tag{18.4}$$

where θ is the angle through which the body has rotated. Since we are now considering rotation of the body about its center of mass, from Eq. (17.7),

$$M = M_c = I_c \ddot{\theta} \tag{18.5}$$

Substitution of Eq. (18.5) into Eq. (18.4) gives

$$W_r = \int I_c \ddot{\theta} \, d\theta$$

$$= \tfrac{1}{2} I_c \omega_B^2 - \tfrac{1}{2} I_c \omega_A^2 \tag{18.6}$$

where ω_A, ω_B are the angular velocities of the body when its center of mass is positioned at A and B, respectively, or when the body has rotated from position A to position B. Thus, the kinetic energy of a body due to rotation abouts its center of mass is given by

$$T_r = \tfrac{1}{2} I_c \omega^2 \tag{18.7}$$

and combining Eqs. (18.6) and (18.7), we get

$$W_r = T_{rB} - T_{rA} = \Delta T_r \tag{18.8}$$

Hence, the work done by a moment about the center of mass of a body is equal to the change in kinetic energy due to rotation about the center of mass.

18.4   Total Kinetic Energy of a Rigid Body

The total kinetic energy possessed by a rigid body can be obtained by adding the kinetic energies due to translation of its center of mass [Eq. (18.2)] and rotation about its center of mass [Eq. (18.7)]. Thus,

$$T = T_t + T_r$$

$$= \tfrac{1}{2} m v_c^2 + \tfrac{1}{2} I_c \omega^2 \tag{18.9}$$

where v_c is the velocity of the center of mass, ω is the angular velocity, and I_c is the moment of inertia of the body about an axis passing through the center of mass.

The validity of Eq. (18.9) can be demonstrated as follows. Figure 18.2 shows one particle of mass m_p in a rigid body. The center of mass C of the

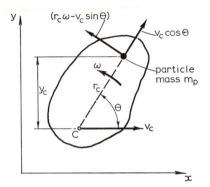

Fig. 18.2 Components of motion of an individual particle in a rigid body.

rigid body is moving with a velocity v_c, and the body has an angular velocity ω. The components of the absolute velocity of the particle are shown in the figure and can be added vectorially to give the absolute velocity v of the particle. Thus,

$$v^2 = (r_c\omega - v_c \sin \theta)^2 + (v_c \cos \theta)^2$$
$$= v_c^2 + r_c^2\omega^2 - 2r_c v_c\omega \sin \theta \qquad (18.10)$$

The kinetic energy T_p of the particle is given by

$$T_p = \tfrac{1}{2}m_p v^2 \qquad (18.11)$$

and substitution of Eq. (18.10) into Eq. (18.11) and summing for all particles give

$$T = \sum T_p = \frac{v_c^2}{2} \sum m_p + \frac{\omega^2}{2} \sum (m_p r_c^2) + v_c\omega \sum (m_p r_c \sin \theta) \quad (18.12)$$

The first term on the right-hand side of Eq. (18.12) is equal to $\tfrac{1}{2}mv_c^2$, and the second term is equal to $\tfrac{1}{2}I_c\omega^2$. Since $r_c \sin \theta$ is equal to y_c and $\sum(m_p y_c)$ is zero, the third term is zero. Hence, Eq. (18.12) becomes

$$T = \tfrac{1}{2}mv_c^2 + \tfrac{1}{2}I_c\omega^2 \qquad (18.13)$$

which is identical to Eq. (18.9).

Finally, Fig. 18.3 shows a rigid body rotating about a fixed axis O. In this case, the velocity of the center of mass is given by

$$v_c = r_o\omega \qquad (18.14)$$

Substitution of Eq. (18.14) into Eq. (18.13) gives

$$T = \tfrac{1}{2}mr_o^2\omega^2 + \tfrac{1}{2}I_c\omega^2$$

and since the moment of inertia of the body about O is given by

$$I_o = I_c + mr_0^2 \qquad (18.15)$$

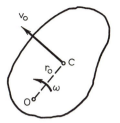

Fig. 18.3   Rigid body rotating about O.

we get

$$T = \tfrac{1}{2}I_o\omega^2 \tag{18.16}$$

18.5   Power

As defined earlier, power is the rate at which work is done. Thus, for a particle, power is given by

$$P = \dot{W} = Fv \tag{18.17}$$

where v is the velocity of the particle and F is the component of the applied force in the direction of the velocity. For a rigid body, Eq. (18.17) still applies. However, in this case, v is the velocity of the point of application of F. If a rigid body is acted upon by a moment M and is rotating at an angular velocity ω, then power can also be expressed by

$$P = \dot{W} = M\omega \tag{18.18}$$

18.6   Potential Energy

During the rotation of a rigid body about its center of mass, the net work done by gravity on all the individual particles is zero, and therefore the change in potential energy due to lowering or raising a rigid body can be calculated by assuming that the mass of the body is concentrated at its center of mass. We can now apply Eq. (15.29) to the solution of problems concerning rigid bodies, remembering that the total kinetic energy is given by Eq. (18.13). Thus,

$$W_n = T_B - T_A + V_B - V_A \tag{18.19}$$

where W_n is the work done by the nonconservative forces.

Example 18.1   A thin disk of radius r is released from rest and allowed to roll down an inclined plane. If friction is sufficient to prevent slipping, deter-

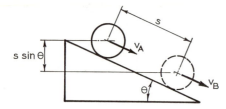

Fig. 18.4   Diagram for Example 18.1.

mine the velocity of the center of the disk after it has rolled a distance s measured down the plane (Fig. 18.4).

Solution.   Since no slipping takes place, the friction force does no work, and since there are no other external forces involved, W_n is zero, and Eq. (18.19) becomes

$$T_A + V_A = T_B + V_B \qquad (a)$$

In this problem, if V_B is taken as zero, then

$$V_A = mgs \sin \theta \qquad (b)$$

From Eq. (18.13),

$$T_A = 0 \qquad (c)$$

and

$$T_B = \tfrac{1}{2}mv_B^2 + \tfrac{1}{2}I_c\omega_B^2 \qquad (d)$$

For rolling without slipping, ω_B is given by v_B/r, and for a disk, I_c is given by $\tfrac{1}{2}mr^2$. Thus, Eq. (d) becomes

$$T_B = \tfrac{3}{4}mv_B^2 \qquad (e)$$

Substitution of Eqs. (b), (c), and (e) into Eq. (a) gives

$$mgs \sin \theta = \tfrac{3}{4}mv_B^2$$

or

$$v_B = (\tfrac{4}{3}gs \sin \theta)^{1/2}$$

Example 18.2   Figure 18.5 shows a block of mass 5 kg connected by an inextensible cable to a windlass whose mass is 35 kg and whose moment of inertia is 3.25 kg·m². If a 50-N force is applied to the lever at A, at a time when the suspended mass is falling at 5 m/s, so that the brake shoe at B is applied, how far will the suspended mass drop before stopping? Assume that the coefficient of friction between the windlass and the brake shoe is 0.5.

Solution.   When the brake lever is applied, the normal force F_n applied to the windlass can be found by taking moments about the lever support O. Thus,

$$100F_n = 200(50)$$

$$F_n = 100 \text{ N} \qquad (a)$$

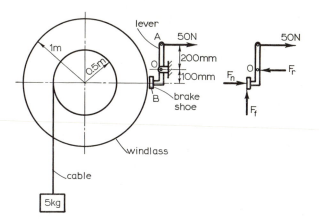

Fig. 18.5 Diagrams for Example 18.2.

Since relative motion between the brake shoe and the windlass will occur until the mass stops falling,

$$F_f = \mu F_n$$
$$= 0.5(100)$$
$$= 50 \text{ N} \tag{b}$$

If the suspended mass travels a distance s, the surface of the windlass, which is traveling at twice the speed of the mass, will have traveled a distance $2s$ relative to the brake shoe. Hence, the work done by the friction force F_f will be

$$W_f = -50(2s) = -100s \tag{c}$$

The total kinetic energy possessed by the system, before application of the brake, will be the sum of the kinetic energy of the windlass and the kinetic energy of the mass. Hence,

$$T_A = \tfrac{1}{2}mv_A^2 + \tfrac{1}{2}I_c\omega_A^2 \tag{d}$$

Since v_A is 5 m/s and ω_A is 10 rad/s (given by v_A divided by the radius of the axle), we get

$$T_A = \tfrac{1}{2}(5)(5)^2 + \tfrac{1}{2}(3.25 \times 10^2)$$
$$= 225 \text{ kg·m}^2/\text{s} \tag{e}$$

The change in potential energy of the mass is given by

$$V_A - V_B = mgs$$
$$= 5(9.81)s$$
$$= 49.05s \tag{f}$$

Finally, when the mass stops moving, the kinetic energy T_B of the system becomes zero, and thus, from Eq. (18.19), we get

$$W_f = T_B - T_A + V_B - V_A$$
$$-100s = -225 + 0 - 49.05s$$

and $\qquad\qquad s = 4.42 \text{ m}$

Example 18.3 Figure 18.6 shows a belt which transmits 25 kW of power to a pulley 500 mm in diameter that rotates at 60 rad/s. Determine the difference in tensions $(F_1 - F_2)$ between the tight side and the slack side of the belt.

Solution. The power P transmitted is given by Eq. (18.18):

$$P = M\omega \qquad\qquad\text{(a)}$$

Thus,

$$25 \times 10^3 = M(60)$$
$$M = 416.7 \text{ N·m} \qquad\qquad\text{(b)}$$

The moment transmitted to the pulley is given by

$$M = 0.25(F_1 - F_2) \qquad\qquad\text{(c)}$$

Hence,

$$F_1 - F_2 = \frac{2}{0.5}(416.7) = 1.7 \text{ kN} \qquad\qquad\text{(d)}$$

Example 18.4 At 25 m/s, an automobile encounters an air resistance of 1 kN. Determine the power delivered to the rear wheels of the automobile if its acceleration is 1 m/s² and its mass is 1900 kg. Neglect the rotational inertia of the wheels and drive train.

Solution. The free-body diagrams for the automobile and rear wheels are shown in Fig. 18.7. It can be seen that

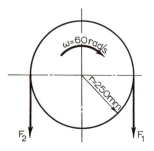

F_2 $\qquad$ F_1 $\qquad$ **Fig. 18.6** Diagram for Example 18.3.

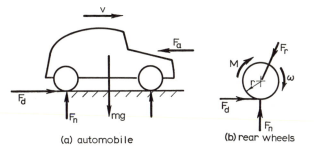

Fig. 18.7 Diagrams for Example 18.4.

$$F_d - F_a = m\ddot{x} \tag{a}$$

where F_d is the total driving force at the rear wheels and F_a is the force due to air resistance. Thus,

$$F_d = F_a + m\ddot{x} \tag{b}$$

The power required at the rear wheels is determined by the rate at which the force F_d is doing work. Thus,

$$P = M\omega = F_d v = (m\ddot{x} + F_a)v$$
$$= [1900(1) + 100]25 = 72.5 \text{ kW} \tag{c}$$

Problems

18.1 Referring to problem 17.21, use energy methods to find the velocity of the cylinder after it has rolled a distance of 0.5 m down the rails.

18.2 A rod AB of length $2l$ is freely pinned at A and arranged in a vertical position with B above A. The rod is displaced slightly so that it swings downward. Derive an expression for its angular velocity when B is directly below A.

18.3 A vertical rod 2 m in length is resting on a horizontal frictionless surface. If it is displaced slightly from rest, determine its angular velocity when it falls completely to the surface.

18.4 An electric motor developing 5 kW and rotating at a frequency of 30 s^{-1} is mounted on four springs each of stiffness 10 kN/m as shown in Fig. P18.4. If the motor casing deflects through the angle θ, determine the magnitude of θ and the direction of rotation of the rotor.

18.5 A thin hollow cylinder of radius r has a particle of equal mass attached symmetrically to its inner surface. If the system is disturbed from its position of stable equilibrium on a rough horizontal plane and then left to itself, show that when the radius to the particle makes an angle θ with the downward vertical, $r\dot{\theta}^2(2 - \cos\theta) - g\cos\theta = $ constant.

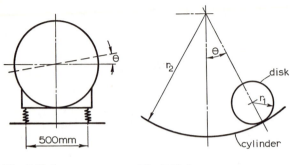

Fig. P18.4   **Fig. P18.6**

18.6 A homogeneous disk of radus r_1 is free to roll on the inner surface of a larger cylinder of radius r_2 (Fig. P18.6). If the disk starts from rest in the position shown, obtain (a) an expression for the velocity of its center of mass when it reaches the bottom of the cylinder and (b) an expression for the normal reaction between the disk and the cylinder when it reaches the bottom of the cylinder.

18.7 Figure P18.7 shows two gears A and B of masses m_1 and m_2, respectively, connected by a rigid link C of mass m. If gear A is initially at rest when a constant torque of magnitude M is applied to the link, obtain an expression for the angular velocity of gear A in terms of the angular displacement θ of the link measured from the vertical. Gear B is fixed.

18.8 A drum 0.3 m in diameter and weighing 400 N is used to lower an anchor weighing 10 kN (Fig. P18.8). The 30-m chain which connects the anchor to the drum has a mass per unit length of 10 kg/m. If the anchor chain has an initial unwrapped length of 1 m when the drum is released from rest, determine the angular velocity of the drum when the anchor has fallen an additional 20 m. Neglect the buoyancy of the anchor and friction in the bearings.

18.9 Figure P18.9 shows a type of slider-crank mechanism. The weight of each uniform link is 50 N, the weight of the slider is 25 N, and the applied torque M is 3 N·m. If the mechanism starts from rest when θ is 45°, determine the angular velocity of each link when θ becomes zero.

18.10 A train is traveling at 5 m/s up an incline of 1 in 10 (1 unit vertically to 10 units along the incline) when the caboose becomes disengaged. How many seconds will elapse before the caboose repasses the point of disengagement on its way down the incline? Assume a constant total resistance to motion for the caboose of 200 N. The caboose has a total weight of 60 kN, and each of its eight wheels weighs 300 N, has a rolling radius of 250 mm, and has a radius of gyration of 200 mm about the axle centerline.

18.11 A uniform circular disk of mass $2m$ and radius r has a particle of mass m fixed to its circumference. The disk is projected with its plane vertical and

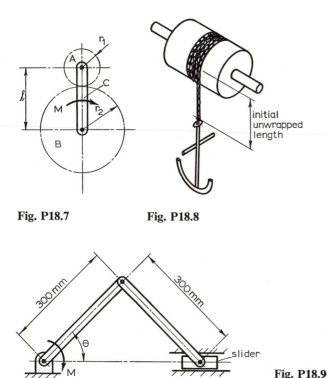

Fig. P18.7   **Fig. P18.8**

Fig. P18.9

the particle initially in its highest positions, so as to roll without slipping on a horizontal rail. Prove that when the radius to the particle makes an angle θ with the upward vertical, $r\dot{\theta}^2 = [7r\omega^2 + 2g(1 - \cos\theta)]/(5 + 2\cos\theta)$, where ω is the angular speed of projection.

18.12   Two uniform rods OA and AB each of mass m and length $2l$ are smoothly pin-jointed at A, and the end O is smoothly pin-jointed to a fixed point on a horizontal table. Initially the rods are at rest and in line on the table. Rod AB is then given an angular velocity $(3g/l)^{1/2}$ about A in a vertical plane. Prove that as long as rod OA remains on the table the angular velocity of rod AB is $(2 - \sin\theta)^{1/2}(3g/2l)^{1/2}$ when it has turned through an angle θ. Find the vertical component of the reaction of OA on AB, and hence find the value of θ at which rod OA will begin to rise.

19

Momentum Applied to the Dynamics of Rigid Bodies

19.1 Introduction

In this final chapter dealing with the principles of dynamics, we shall see how the momentum equations for particles, developed in Chapter 16, can be applied to rigid bodies. We shall also consider situations involving impact forces acting on rigid bodies.

19.2 Momentum

The translational and rotational equations of motion for plane rigid body motion are given by

$$F_x = ma_{cx}, \quad F_y = ma_{cy}, \quad M = I\alpha \qquad (19.1)$$

where a_{cx} and a_{cy} are the components of acceleration of the center of mass of the body, m is the total mass of the body, α is the angular acceleration of the body, and I is the moment of inertia of the body about either its center of mass or about a fixed point. These equations can also be written as

$$F_x = \frac{d}{dt}(mv_{cx}), \quad F_y = \frac{d}{dt}(mv_{cy}), \quad M = \frac{d}{dt}(I\omega) \qquad (19.2)$$

where mv_{cx} and mv_{cy} are the components of linear momentum of the body and $I\omega$ is the angular momentum of the body about either a fixed point or the center of mass. To illustrate that $I\omega$ is the angular momentum of a rigid body

278

we shall refer to Fig. 19.1, which shows the motion of one particle of mass m_p in a rigid body. Since angular momentum is the moment of linear momentum, from Fig. 19.1a, the total angular momentum for all particles about the center of mass C of the body is

$$p_c = (\sum m_p r_c^2)\omega + (\sum m_p r_c \cos \theta)v_c \tag{19.3}$$

and since $\sum m_p r_c \cos \theta$ is zero,

$$p_c = (\sum m_p r_c^2)\omega = I_c \omega \tag{19.4}$$

From Fig. 19.1b, the angular momentum about the fixed point O is given by

$$p_o = (\sum m_p r_o^2)\omega = I_o \omega \tag{19.5}$$

Thus, the rotational equation of motion about the center of mass can be written as

$$M_c = \frac{d}{dt}(I_c \omega) \tag{19.6}$$

while about a fixed point,

$$M_o = \frac{d}{dt}(I_o \omega) \tag{19.7}$$

Integrating Eqs. (19.2), we get

$$\int_{t_1}^{t_2} F_x \, dt = mv_{cx2} - mv_{cx1} \tag{19.8}$$

$$\int_{t_1}^{t_2} F_y \, dt = mv_{cy2} - mv_{cy1} \tag{19.9}$$

$$\int_{t_1}^{t_2} M \, dt = I\omega_2 - I\omega_1 \tag{19.10}$$

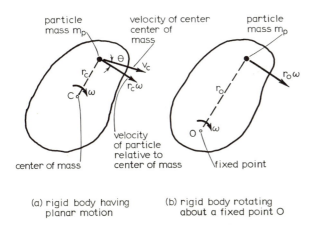

particle mass m_p

velocity of center center of mass

particle mass m_p

center of mass

velocity of particle relative to center of mass

fixed point

(a) rigid body having planar motion

(b) rigid body rotating about a fixed point O

Fig. 19.1 Motion of one particle in a rigid body.

where M and I are the moment and moment of inertia, respectively, about either the center of mass or a fixed point. It follows from these equations that in a system of interacting bodies where the forces due to collisions, etc., are always equal and opposite and where no external forces or moments are acting (i.e., $\int F_x \, dt = \int F_y \, dt = \int M \, dt = 0$),

$$mv_{cx2} = mv_{cx1} \tag{19.11}$$

$$mv_{cy2} = mv_{cy1} \tag{19.12}$$

Hence, linear momentum is conserved. Also,

$$I\omega_2 = I\omega_1 \tag{19.13}$$

and angular momentum is conserved. Thus, as with particles, if the resultant force acting on a rigid body is zero, then the linear momentum of the body must be conserved. In addition, if the moment acting on a body is zero, then the angular momentum must be conserved.

19.3 Impulse and Impact

If the time interval, over which an external force or moment is acting, approaches zero, then we get the following equations for impact forces and moments:

$$\hat{F}_x = m(v'_{cx} - v_{cx}) \tag{19.14}$$
$$\hat{F}_y = m(v'_{cy} - v_{cy}) \tag{19.15}$$
$$\hat{M}_c = I_c(\omega' - \omega) \tag{19.16}$$
$$\hat{M}_o = I_o(\omega' - \omega) \tag{19.17}$$

where $\hat{F}_x$ represents $\int F_x \, dt$ when $\Delta t \to 0$, etc., and where the prime represents the linear or angular velocity after impact.

If a system consists of a series of interconnected rigid bodies, then the equations of motion can be written for each individual body and summed. Since the reaction forces and moments at the points where the bodies are connected are equal and opposite to each other, these reaction terms will cancel one another and will not appear in the force and moment summations. The resulting equations of motion are

$$F_x = \sum \frac{d}{dt}(mv_{cx}) \tag{19.18}$$

$$F_y = \sum \frac{d}{dt}(mv_{cy}) \tag{19.19}$$

$$M_c = \sum \frac{d}{dt}(I_c\omega) \tag{19.20}$$

$$M_o = \sum \frac{d}{dt}(I_o\omega) \tag{19.21}$$

Thus, for a given direction, the component of the resultant external force is equal to the sum of the rates of change of linear momenta for all the bodies. Also, for rotation about the center of mass or about a fixed point, the moment of the external forces is equal to the sum of the rates of change of angular momenta.

Example 19.1 A uniform rod of length l is suspended vertically from a pin joint at one end as shown in Fig. 19.2a. If the rod is struck with an impulse $\hat{F}$ at a distance b from the pin joint, determine the impulse reactions at the pin joint.

Solution. Employing Eq. (19.17) and referring to Fig. 19.2b, we get

$$\hat{M}_o = \hat{F}b = I_o\omega' - I_o\omega \tag{a}$$

where $I_o\omega'$ and $I_o\omega$ are the angular momenta of the body immediately before and after impact. Thus, since the rod is initially at rest, ω is zero, and

$$\hat{F}b = I_o\omega' = \tfrac{1}{3}ml^2\omega' \tag{b}$$

or $$\omega' = \frac{3\hat{F}b}{ml^2} \tag{c}$$

Resolving horizontally, we get

$$\hat{F} - \hat{F}_h = m(v' - v) \tag{d}$$

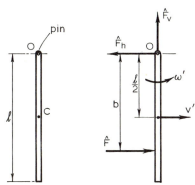

(a) before impact   (b) after impact   **Fig. 19.2** Diagrams for Example 19.1.

and since v is zero and v' is $\omega'l/2$, we get

$$\hat{F}_h = \hat{F} - \frac{ml\omega'}{2} \qquad (e)$$

Substitution of Eq. (c) into Eq. (e) gives

$$\hat{F}_h = \hat{F}\left(1 - \frac{3}{2}\frac{b}{l}\right) \qquad (f)$$

Resolving vertically, we get

$$\hat{F}_v = 0 \qquad (g)$$

It is interesting to consider the condition where the impulse reaction at the pin joint is zero. In this case, from Eq. (f),

$$\frac{3}{2}\frac{b}{l} - 1 = 0$$

or $$b = \tfrac{2}{3}l \qquad (h)$$

The point defined by Eq. (h) is known as the center of percussion.

Example 19.2 A small merry-go-round intended for preschool children consists of a turntable which is driven with a constant angular velocity ω (Fig. 19.3). If the attendant of mass m moves radially outward with a velocity v relative to the turntable, obtain an expression for the torque (moment) required to rotate the turntable as a function of the attendant's position.

Solution.   Taking moments about the vertical axis O [Eq. (19.21)], we get

$$M_o = \sum \frac{d}{dt}(I_o\omega) \qquad (a)$$

or $$M_o = \frac{d}{dt}(I_{ot}\omega + mr^2\omega) \qquad (b)$$

where I_{ot} is the moment of inertia of the turntable about its axis. From Eq. (b) we get

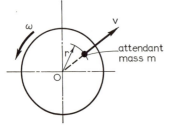

Fig. 19.3   Diagram for Example 19.2.

$$M_o = 2mr\omega\frac{dr}{dt} = 2mr\omega v \qquad (c)$$

Example 19.3   A 20-mm cube is projected along a horizontal surface, and it encounters a small step (Fig. 19.4a). What is the minimum velocity required for the cube to rotate about the edge of the step and just turn over? Assume plastic impact between the cube and the step.

Solution.   With plastic impact between the part and the step at A, the part rotates about A after impact. It is first necessary to determine the angular velocity ω' of the part immediately after impact in terms of its initial velocity v. One way of doing this is to take the sum of the moments about the center of mass of the part and to equate the net moment to the change in angular momentum about its center of mass. Thus,

$$\hat{M}_c = p'_c - p_c \qquad (a)$$

where p_c and p'_c represent the angular momenta immediately before and after impact, respectively. Since p_c is zero, from Fig. 19.4b,

$$\hat{M}_c = \hat{F}_h a - \hat{F}_v a = I_c\omega' \qquad (b)$$

where $2a$ is the length of a side of the cubic part. Resolving horizontally gives

$$\hat{F}_h = m(v - a\omega') \qquad (c)$$

and resolving vertically gives

$$\hat{F}_v = ma\omega' \qquad (d)$$

Substitution of Eqs. (c) and (d) into Eq. (b) and writing $I_c = 2ma^2/3$ gives, after rearrangement,

$$\omega' = \frac{3v}{8a} \qquad (e)$$

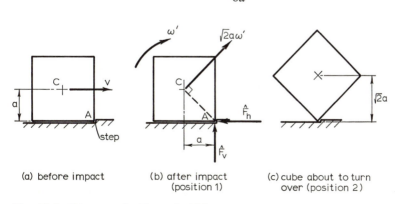

(a) before impact   (b) after impact   (c) cube about to turn
                                     (position 1)   over (position 2)

Fig. 19.4   Diagrams for Example 19.3.

After impact, the part continues to rotate about A. For the part to just turn over, its center of mass must reach a position vertically above A, as shown in Fig. 19.4c. For the conservation of energy between positions 1 and 2, we can write

$$T_1 + V_1 = T_2 + V_2$$

or

$$\tfrac{1}{2}I_A(\omega')^2 = mg(\sqrt{2}a - a) \tag{f}$$

Now

$$I_A = I_c + m(\sqrt{2}a)^2 = \frac{2}{3}ma^2 + 2ma^2 = \frac{8ma^2}{3} \tag{g}$$

and substituting Eqs. (e) and (g) into Eq. (f), we get, after rearrangement,

$$v^2 = \frac{16}{3}(\sqrt{2} - 1)ag = 2.2ag = 0.22$$

or

$$v = 0.47 \text{ m/s} = 470 \text{ mm/s}$$

Alternative Method of Solution. In writing the impact equations, a procedure similar to that described for writing equations of motion for a rigid body can be adopted. If two free-body diagrams are drawn, one showing the impulse forces and the other showing two components of the change in momentum of the center of mass together with the change in angular momentum about the center of mass, the two diagrams can be equated and moments taken about any point in the diagrams. These diagrams are shown in Fig. 19.5 for the present example. Taking moments about A, we get

$$0 = ma(a\omega' - v) + ma^2\omega' + I_c\omega' \tag{h}$$

Substitution of $I_c = 2ma^2/3$ and rearrangement give

$$\omega' = \frac{3v}{8a} \tag{i}$$

which is identical to Eq. (e).

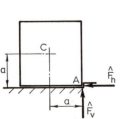

(a) impulse forces

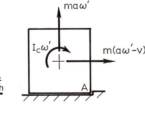

(b) change in momentum

Fig. 19.5 Diagrams for Example 19.3 (alternative method).

Problems

19.1 A 50-kN crate (Fig. P19.1), whose radius of gyration is 3 m about an axis passing through its center of mass and perpendicular to the page, falls while being unloaded from a cargo ship. If the velocity of the crate just before impact is 12 m/s and if the crate is slightly inclined when one edge hits the dock and does not rebound, determine the angular velocity of the crate immediately after impact.

19.2 A disk of radius r is rolling without slipping on a rough horizontal surface when it encounters an inclined surface (Fig. P19.2). Obtain an expression for the ratio of the velocities of the disk immediately before and after its impact with the inclined surface, assuming that the disk rolls up the inclined surface. In addition, obtain an expression for the energy loss that occurs during impact.

19.3 If the disk described in problem 19.2 encounters a small step of height h, obtain an expression for the minimum initial velocity v_0 for the disk to mount the step.

19.4 A ballistic pendulum has a mass m_1 and a moment of inertia I_0 about the fixed point O (Fig. P19.4). If a bullet of mass m_2 is fired into the pendulum as shown in the figure and the pendulum rotates through an angle θ, obtain an expression for the initial velocity of the bullet.

19.5 A 10-kg man, 2 m in height, is wearing ice skates and is standing in the middle of a frozen lake. While holding a rifle horizontally against his shoulder, which is 650 mm above his center of mass, he discharges the rifle, firing an 0.06-kg bullet at 500 m/s. Determine the man's velocity immediately after the rifle discharges. Neglect the friction between the skates and the ice and the mass of the rifle. The man's center of mass is 1 m above the ice, and his moment of inertia about a horizontal axis through his center of mass is 3 kg·m².

19.6 Figure P19.6 shows four small spheres, each weighing 1.5 N and held by light strings in four slots in a circular disk of weight 60 N and radius of gyration about its axis of 200 mm. If the strings are cut while the disk is rotating and the spheres come to rest at the ends of the slots, find the new angular velocity of the disk. Also, find the magnitude of the energy loss due to the impact of the spheres with the ends of the slots. Treat the spheres as point masses.

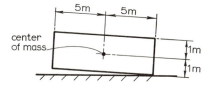

Fig. P19.1

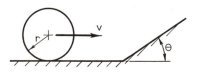

Fig. P19.2

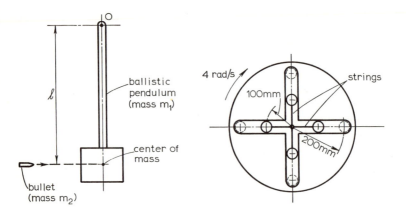

Fig. P19.4   **Fig. P19.6**

19.7 Two pulleys A and B initially rotate about different parallel axes in opposite directions at $3\,s^{-1}$ and $5\,s^{-1}$, respectively. The diameters of pulleys A and B are 0.8 m and 0.4 m, respectively, and their moments of inertia are $5\,kg\cdot m^2$ and $1.2\,kg\cdot m^2$, respectively. The pulleys are suddenly connected by a belt which passes around both pulleys and which is simultaneously tightened. Calculate the speed of pulley A and its direction of rotation after the belt has stopped slipping.

19.8 A uniform rod OA of mass m and length $2a$ turns freely in a horizontal plane about a fixed vertical axis through O. A similar rod AB is hinged to the first at A so that it turns in a horizontal plane. At the moment when OA and AB are in line and rotating with respective angular velocities ω_0 and ω_b in opposite senses, the hinge at A is suddenly locked. If the rods come to rest, prove that $5\omega_b = 11\omega_0$, and find the impulsive reaction on the axis through O in terms of ω_0.

19.9 Two equal uniform rods AB and BC each of mass m and length l are pin-jointed together at B. When the rods rest on a smooth horizontal table with AB and BC in line, C is jerked into motion with a velocity v perpendicular to the rods. If ABC remains a straight line after motion begins, prove that the friction at the joint must supply an impulsive couple of magnitude $mvl/8$.

19.10 A lamina is rotating with an angular velocity ω about an axis through its mass center perpendicular to its plane when some other point of the lamina is suddenly fixed. Show that the kinetic energy is reduced in the same ratio as the angular velocity. If the lamina is a uniform square of side $2l$ and mass m and it is found that half the kinetic energy is lost, find the distance from the mass center to the point which is pinned and the impulse on the pin.

20

Some Further Applications

20.1 Geared Systems

To analyze the accelerations of geared systems, it is necessary to determine their effective moment of inertia. Figure 20.1 shows two shafts A and B connected by gearing so that shaft B rotates at N times the speed of rotation of shaft A. Let I_A and I_B be the moments of inertia about the shaft axes of gears A and B, respectively, and let α_A be the angular acceleration of shaft A. To find the torque or moment M_A required on shaft A to produce this acceleration we proceed as follows:

(1) Since the rotational motions are in the ratio N, the angular acceleration of B is

$$\alpha_B = N\alpha_A \qquad (20.1)$$

(2) Taking moments about shaft B, we get

$$F_y r_B = I_B \alpha_B \qquad (20.2)$$

(3) Taking moments about shaft A, we get

$$M_A = I_A \alpha_A + F_y r_A \qquad (20.3)$$

Finally, substitution of Eqs. (20.1) and (20.2) into Eq. (20.3) gives, after rearrangement,

$$M_A = \alpha_A (I_A + I_B N^2) \qquad (20.4)$$

The term in parentheses can be regarded as the equivalent moment of inertia of the system referred to shaft A.

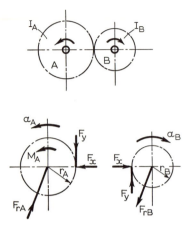

Fig. 20.1   Geared shafts.

The above analysis can be extended to a system where there are several shafts geared together. In this case, the equivalent moment of inertia I_e, referred to shaft A, would be given by

$$I_e = I_A + \sum I_x N_x^2 \tag{20.5}$$

where I_x is the moment of inertia of the xth shaft and N_x is the ratio of the rotational speed of the xth shaft to the rotational speed of shaft A.

Example 20.1   A motorcycle and rider together weigh 1.6 kN. With an engine speed of 30 s^{-1} and a 9 to 1 gear reduction to the back wheel, the engine exerts a torque of 20 N·m. The moment of inertia of each wheel (I_w) is 1.2 kg·m^2 and of the engine parts (I_e) is 0.08 kg·m^2. The total resistance to motion (wind, friction, etc.) is 80 N, and the effective diameter of the road wheels is 0.64 m. Find the velocity of the motorcycle and its acceleration under these conditions.

Solution.   The engine speed is 30 s^{-1}, and since the gear reduction is 9 to 1, the angular velocity ω of the back wheel will be $\frac{30}{9}$ s^{-1}. The velocity v of the motorcycle is given by

$$v = r\omega \tag{a}$$

where r is the radius of the back wheel. Thus,

$$v = \frac{0.64}{2}\left(\frac{30}{9}\right) = 1.07 \text{ m/s}$$

If α_e is the angular acceleration of the engine, then the angular acceleration of the wheels α_w will be

$$\alpha_w = \frac{\alpha_e}{9} \tag{b}$$

and the linear acceleration a of the motorcycle will be

$$a = r\alpha_w \tag{c}$$

Figure 20.2a shows a free-body diagram for the motorcycle and rider. Resolving horizontally, we get

$$F_{f1} - F_{f2} - F_r = ma \tag{d}$$

where F_{f1} is the frictional force at the back wheel, F_{f2} is the frictional force at the front wheel, and F_r is the force resisting motion. Taking moments about the front-wheel axle for the front wheel only (Fig. 20.2b) gives

$$F_{f2}r = I_w\alpha_w$$

or

$$F_{f2} = \frac{I_w\alpha_w}{r} = \frac{1.2\alpha_w}{0.32} = 3.75\alpha_w \tag{e}$$

and since $F_r = 80$ N and $mg = 1.6$ kN, we get, from Eqs. (c), (d), and (e),

$$F_{f1} = ma + F_{f2} + F_r$$

$$= \frac{1600}{9.81}(0.32)\alpha_w + 3.75\alpha_w + 80$$

$$= 56\alpha_w + 80 \tag{f}$$

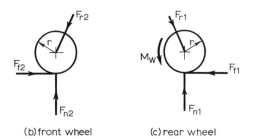

(a) motorcycle and rider

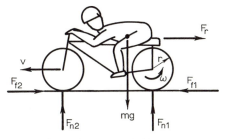

(b) front wheel   (c) rear wheel

Fig. 20.2 Free-body diagrams for Example 20.1.

The torque M_w on the rear wheel must provide this force together with the torque to accelerate the rear wheel. Thus, referring to the free-body diagram for the rear wheel (Fig. 20.2c) and taking moments about the axle, we get

$$M_w = r(56\alpha_w + 80) + I_w\alpha_w$$
$$= 0.32(56\alpha_w + 80) + 1.2\alpha_w$$
$$= 19.1\alpha_w + 25.6 \tag{g}$$

The torque M_e at the engine must provide torque to accelerate the engine and the torque M_w at the rear wheel. Thus,

$$M_e = I_e\alpha_e + \frac{1}{9}M_w \tag{h}$$

and since

$$\alpha_e = 9\alpha_w \tag{i}$$

we get

$$M_e = 9I_e\alpha_w + \frac{1}{9}(19.1\alpha_w + 25.6)$$

Thus,

$$20 = 9(0.08)\alpha_w + \frac{1}{9}(19.1\alpha_w + 25.6)$$

and

$$\alpha_w = 6.04 \text{ rad/s}^2$$

Finally, from Eq. (c),

$$a = r\alpha_w$$
$$= 0.32(6.04) = 1.93 \text{ m/s}^2$$

20.2 Balancing of Rotating Masses

If the center of mass of a rotating body does not lie on the axis of rotation, then its mass may be regarded as being concentrated at a single point a distance r from the axis of rotation. For example, Fig. 20.3a shows a disk which is supported eccentrically on a shaft. The equivalent point-mass system is shown in Fig. 20.3b. If the system is stationary, the weight of the mass causes vertical reactions at the bearings (Fig. 20.3c). As the system rotates, a rotating inertia force causes rotating dynamic reactions at the bearings (Fig. 20.3d) which are superimposed on the static reactions due to the weight of the mass. For the purpose of balancing a system, consideration need only be given to the dynamic portion of the reactions, and the static portion can be ignored since it is always present and remains constant in magnitude and direction.

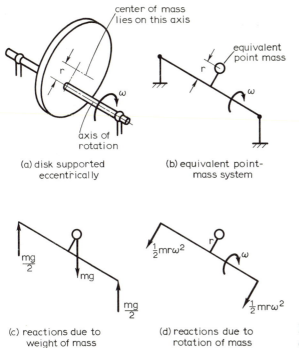

(a) disk supported
eccentrically

(b) equivalent point-
mass system

(c) reactions due to
weight of mass

(d) reactions due to
rotation of mass

Fig. 20.3 Rotating
eccentric mass.

If the angular velocity of the rotating system is ω, the inertia force giving rise to dynamic reactions at the supports has a magnitude of $mr\omega^2$. This inertia force can be reduced to zero if adjustments can be made to the system so that the center of mass of the system lies on its axis of rotation. It should be noted that the inertia force introduced is proportional to the square of the angular velocity, and for high-speed machinery, this inertia force can be very much greater than the weight of the mass.

The center of mass of a system may not be on the axis of rotation (1) due to the geometry of the system (e.g., a crankshaft) or (2) due to the lack of homogeneity in the material (e.g., blow holes, slag inclusions, etc.). So, even in the case of a perfectly round cylinder rotating about its geometrical axis, it is necessary to perform an experiment to confirm that the center of mass lies on the geometric axis if the cylinder is to be used in high-speed machinery.

20.2.1 Single-Mass Systems

If we consider inertia forces only, then because of the eccentricity of the mass (Fig. 20.4), there will be an inertia force of magnitude $mr\omega^2$, which, for constant angular velocity, is constant in magnitude but variable in direction. A balancing inertia force can be produced by attaching a balancing mass m_b to the shaft at radius r_b so that $m_b r_b \omega^2$ is equal to $mr\omega^2$; i.e.,

$$m_b r_b = mr \tag{20.6}$$

To balance a single-mass system by the addition of one other mass, it is essential that the planes of rotation of both masses coincide; otherwise, although the combined center of mass will lie on the axis of rotation, an inertia couple will be introduced which will in itself produce rotating dynamic reactions at the supports (Fig. 20.5). In this case, the system is said to be statically balanced but not dynamically balanced.

20.2.2 Multimass, Uniplanar Systems

Referring to Fig. 20.6a, to find the location and magnitude of the balancing mass m_b, it is necessary to find the equilibrant of the three unbalanced inertia

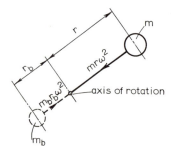

Fig. 20.4 Balancing of single-mass system.

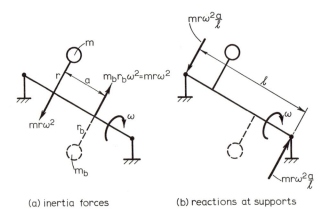

(a) inertia forces   (b) reactions at supports

Fig. 20.5 System statically balanced but not dynamically balanced.

forces $m_1 r_1 \omega^2$, $m_2 r_2 \omega^2$, and $m_3 r_3 \omega^2$. This is illustrated in the vector diagram in Fig. 20.6b. Since the term ω^2 is common throughout, it is only necessary to consider the terms $m_1 r_1$, $m_2 r_2$, etc., in drawing the vector diagram or in analyzing the problem. Analytically, we can say for equilibrium, resolving horizontally,

$$m_1 r_1 \cos \theta_1 + m_2 r_2 \cos \theta_2 + m_3 r_3 \cos \theta_3 + m_b r_b \cos \theta_b = 0 \quad (20.7)$$

and, resolving vertically,

$$m_1 r_1 \sin \theta_1 + m_2 r_2 \sin \theta_2 + m_3 r_3 \sin \theta_3 + m_b r_b \sin \theta_b = 0 \quad (20.8)$$

Solution of Eqs. (20.7) and (20.8) will give the required values of θ_b and $m_b r_b$.

Example 20.2 A mass of 1 g is to be used to balance the uniplanar system shown in Fig. 20.7a. Determine the location of this mass.

Solution. The vector diagram for the products mr is shown in Fig. 20.7b where it can be seen that θ_b is zero and that

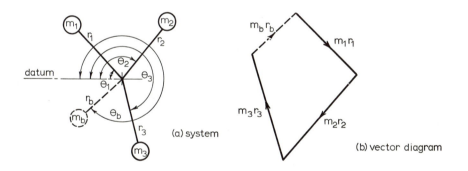

Fig. 20.6 Balancing a multimass, uniplanar system.

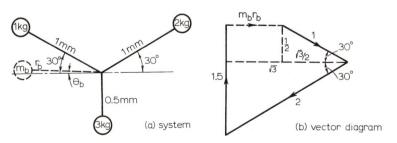

Fig. 20.7 Diagrams for Example 20.2.

$$m_b r_b = \sqrt{3} - \frac{\sqrt{3}}{2} = \frac{\sqrt{3}}{2} = 0.866 \text{ g} \cdot \text{m} \qquad \text{(a)}$$

Since m_b is equal to 1 g, from Eq. (a),

$$r_b = 0.866 \text{ m} \qquad \text{(b)}$$

Alternatively, using Eqs. (20.7) and (20.8), we get

$$1(1) \cos 30° + 2(1) \cos 150° + 3(0.5) \cos 270° + m_b r_b \cos \theta_b = 0 \quad \text{(c)}$$

$$1(1) \sin 30° + 2(1) \sin 150° + 3(0.5) \sin 270° + m_b r_b \sin \theta_b = 0 \quad \text{(d)}$$

From Eq. (c),

$$m_b r_b \cos \theta_b = 0.866 \qquad \text{(e)}$$

and from Eq. (d),

$$m_b r_b \sin \theta_b = 0 \qquad \text{(f)}$$

Hence, from Eq. (f),

$$\theta_b = 0 \qquad \text{(g)}$$

and from Eq. (e),

$$m_b r_b = 0.866 \qquad \text{(h)}$$

20.2.3 Multimass, Multiplanar Systems

Figure 20.8 shows a shaft supported in bearings and carrying two equal masses whose centers of mass are both a distance r from the axis of rotation but on opposite sides of this axis. As explained earlier, since the center of mass of the complete system lies on the axis of rotation, it is statically bal-

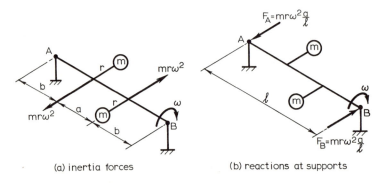

(a) inertia forces   (b) reactions at supports

Fig. 20.8 Dynamically unbalanced system.

anced. However, as the system rotates, a couple is introduced by the inertia forces which produces dynamic rotating reactions at the bearings. Figure 20.8b shows the external forces (reactions), and Fig. 20.8a shows the inertia forces acting when the weights of the masses are neglected. Taking moments about bearing A, we get

$$F_B l = mr\omega^2(a + b) - mr\omega^2(b)$$

or
$$F_B = \frac{a}{l} mr\omega^2 \qquad\qquad (20.9)$$

Similarly, taking moments about bearing B, we get

$$F_A = \frac{a}{l} mr\omega^2 \qquad\qquad (20.10)$$

For dynamic balance, the reactions F_A and F_B, due to rotation of the masses, must be zero. This can be accomplished by providing two additional eccentric masses that will produce an equal and opposite inertia couple to that produced by the original masses. The magnitude of the required balancing couple M_b is equal to $F_A l$, or

$$M_b = amr\omega^2 \qquad\qquad (20.11)$$

Thus, if the masses were of magnitude m_b and arranged on the shaft at radii r_b a distance d apart, as shown in Fig. 20.9, we could write

$$dm_b r_b \omega^2 = amr\omega^2$$

or
$$dm_b r_b = amr \qquad\qquad (20.12)$$

This analysis was for a particular case; we shall now show how a general multimass, multiplanar system can be balanced.

Figure 20.10a shows a single mass m located at a radius r from a shaft at A.

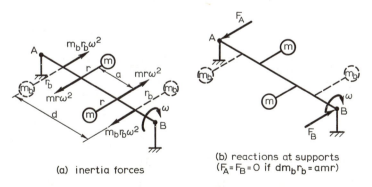

(a) inertia forces

(b) reactions at supports
$(F_A = F_B = 0$ if $dm_b r_b = amr)$

Fig. 20.9   Statically and dynamically balanced system.

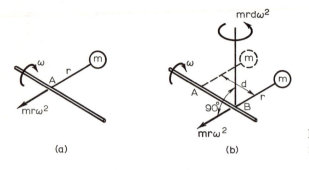

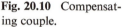

(a)   (b)   **Fig. 20.10** Compensating couple.

As the shaft rotates at a constant angular velocity ω, the inertia force is radial and equal to $mr\omega^2$. If the mass is moved to point B (Fig. 20.10b) on the shaft a distance d from A, then, so that the net forces on the shaft will remain the same, a compensating couple must be introduced. This couple, of magnitude $mrd\omega^2$, can be represented by a vector at right angles to the inertia force vector $mr\omega^2$. Hence, in a problem where several masses are arranged in different planes along a shaft, each mass can be transferred to some convenient reference plane. This will result in a multimass, multiplanar system in that reference plane which may be balanced by the method shown in the previous section. However, for each mass transferred, a compensating couple must be introduced into the system. These couples must also be balanced, and since it is only possible to balance a couple by a couple, the balancing couple must consist of equal and opposite forces in separate planes. Therefore, two planes of reference are required for the balancing masses.

Example 20.3 Balance the system shown in Fig. 20.11 by masses in planes L and M at radii of 6 mm and 8 mm, respectively.

Solution. The masses will all be transferred to the plane M, and the problem can be tabulated as follows:

Plane	m, Mass (kg)	r, Radius (mm)	mr	d, Distance from M	mrd, Compensating Couple
A	4	5	20	26	520
B	5	5	25	18	450
C	6	6	36	8	288
L	3.96	6	23.75	32	760
M	2.31	8	18.5	0	0

Although the vectors representing the compensating couples always lie at right angles to the radius arms of the respective masses, we can draw the vectors for the couples in line with the radius arms. This will simply have the effect of rotating the vector diagram through 90°. Figure 20.12a shows the vector diagram for the couples, and it can be seen that a balancing couple having an *mrd* value of 760 is required in plane *L* at an angle 49° from the horizontal. This gives the angular location of the balancing mass in plane *L* and is shown dashed in Fig. 20.11. Working backwards in the table, the *mr* value for the balancing mass *L* is equal to 760/32 or 23.75, and its mass *m* is equal to 23.75/6 or 3.96 kg.

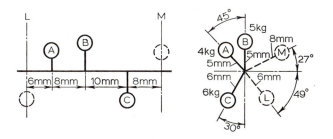

Fig. 20.11 Geometry for Example 20.3.

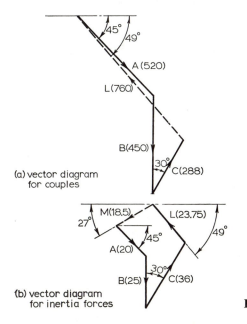

(a) vector diagram
     for couples

(b) vector diagram
     for inertia forces

Fig. 20.12 Vector diagrams.

To find the balancing mass in plane M, the vector diagram for the inertia forces mr referred to plane M is now drawn (Fig. 20.12b). The balancing inertia force is 18.5 and lies at an angle of 27° from the horizontal. This gives the angular location of the balancing mass M, which is shown dashed in Fig. 20.11. Finally, entering the mr value for M in the table, we find that the mass required is 18.5/8 or 2.31 kg.

20.3 Force-Free Undamped Vibrations

A vibrational or oscillatory motion of a body is referred to as *force-free* or *natural* when the body is acted upon only by the restoring force arising from its displacement. If any frictional resistances act on the body, the motion is said to be *damped*; if an externally applied force acts on the body, the motion is said to be *forced*. In this section, we shall deal only with force-free undamped vibrations where the restoring force is linearly proportional to the displacement of the body. Under these circumstances, the resulting vibrational motion is called simple harmonic motion.

20.3.1 Simple Harmonic Motion

A body moves with simple harmonic motion if its acceleration is proportional to its displacement from a fixed point and is always directed toward that point. Figure 20.13 shows a point P, rotating in a circular path of radius r about a fixed point O, with constant angular velocity ω. We shall consider the motion of point Q, which is the projection of point P onto the diameter AB of the circle. The displacement of Q from its midposition is given by

$$x = r \sin \theta \qquad (20.13)$$

where r is its maximum displacement and is termed the *amplitude* of the

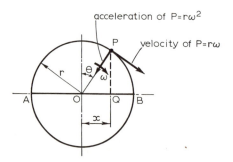

Fig. 20.13   Point P moving in circular path.

motion. The velocity of Q is equal to the component of the velocity of P parallel to AB. Thus,

$$v = r\omega \cos \theta = \omega\sqrt{r^2 - x^2} \qquad (20.14)$$

and the maximum velocity v_{max} is given by

$$v_{max} = r\omega \qquad (20.15)$$

which occurs when x is equal to zero. The acceleration of Q is the component of the acceleration of P parallel to AB. Thus,

$$a = -r\omega^2 \sin \theta = -\omega^2 x \qquad (20.16)$$

and $$a_{max} = r\omega^2 \qquad (20.17)$$

which occurs when x is equal to r. Thus, the acceleration of Q is proportional to its displacement from O and is always directed toward O, so that the motion of Q is simple harmonic. Of course, the above equations could have been obtained by repeated differentiation of Eq. (20.13) with respect to time and realizing that $\dot{\theta}$ is equal to ω.

The periodic time T, for a free vibration, is the time for one cycle. In the present case, this is equal to the time for one revolution of P. Thus,

$$T = \frac{2\pi}{\omega} \qquad (20.18)$$

From Eq. (20.16),

$$\omega = \sqrt{-\frac{a}{x}}$$

or $$T = 2\pi\sqrt{-\frac{x}{a}} \qquad (20.19)$$

Finally, the frequency of vibration n is given by

$$n = \frac{1}{T}$$

and therefore,

$$n = \frac{1}{2\pi}\sqrt{-\frac{a}{x}} \qquad (20.20)$$

20.3.2   Angular Simple Harmonic Motion

The above analysis applies equally to the case of a body having angular oscillations. Thus, if the amplitude of the motion is ϕ and the displacement θ, the angular acceleration is given by

$$\alpha = \theta\omega^2 \tag{20.21}$$

The periodic time is given by

$$T = 2\pi\sqrt{-\frac{\theta}{\alpha}} \tag{20.22}$$

and the frequency is

$$n = \frac{1}{2\pi}\sqrt{-\frac{\alpha}{\theta}} \tag{20.23}$$

20.3.3   Simple Pendulum

Figure 20.14 shows a simple pendulum where a particle of mass m is supported by a light string of length l. If the pendulum is given a small angular displacement θ from the vertical, then taking moments about O, we get

$$-mgl\sin\theta = ml^2\ddot{\theta} = ml^2\alpha$$

or since θ is small, we get

$$\frac{\theta}{\alpha} = -\frac{l}{g} \tag{20.24}$$

Since the ratio θ/α is a negative constant, we have simple harmonic motion, and substitution of Eq. (20.24) into Eq. (20.22) gives the periodic time for small oscillations. Thus,

$$T = 2\pi\sqrt{\frac{l}{g}} \tag{20.25}$$

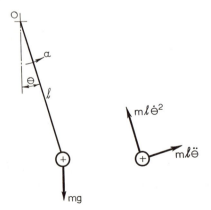

Fig. 20.14   Simple pendulum.

20.3.4   Compound Pendulum

Figure 20.15 shows a rigid body of mass m and radius of gyration k_c about its center of mass, suspended from a point located a distance l from its center of mass. This arrangement is known as a compound pendulum, and for a small displacement θ, taking moments about O, we get

$$-mgl \sin \theta = I_0 \alpha \qquad (20.26)$$

Since θ is small and $I_0 = mk_c^2 + ml^2$, we get

$$\frac{\theta}{\alpha} = -\left(\frac{k_c^2 + l^2}{gl}\right) \qquad (20.27)$$

and from Eq. (20.22), the periodic time is given by

$$T = 2\pi \sqrt{\frac{k_c^2 + l^2}{gl}} \qquad (20.28)$$

The compound pendulum provides an experimental method for the determination of the moment of inertia of a body if the location of its center of mass is known.

Example 20.4   A connecting rod of weight 400 N is suspended on a knife edge through the bore in its small end at a point 100 mm from its center of mass. The period of small oscillations was measured to be 0.71 s. Determine the moment of inertia of the connecting rod about an axis through its center of mass perpendicular to the plane of oscillation.

Solution.   From Eq. (20.28) and since $l = 0.1$ m and $T = 0.71$ s, we get

$$0.71 = 2\pi \sqrt{\frac{k_c^2 + 0.01}{0.1(9.81)}}$$

or

$$k_c = 0.05 \text{ m}$$

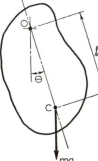

Fig. 20.15   Compound pendulum.

Finally,

$$I_c = mk_c^2$$

$$= \frac{400}{9.81}(0.05)^2$$

$$= 0.1 \ \text{kg} \cdot \text{m}^2$$

20.3.5 Trifilar Suspension

Another method for the experimental determination of the moment of inertia of an object is the use of the trifilar suspension. In this case, a horizontal circular platform is suspended by three equally spaced vertical wires (Fig. 20.16). Measurement of the period of oscillation, with and without the object placed on the platform, allows the moment of inertia of the object to be estimated.

If the platform shown in Fig. 20.16 is displaced through a small angle θ, the corresponding angular displacement ϕ of the wires is given by

$$\phi = \frac{r}{l}\theta \tag{20.29}$$

The tension in each wire is $F_w/3$, and the horizontal component of this tension, tending to restore the platform to its original position, is $(F_w \sin \phi)/3$ or, for small angles, $F_w\phi/3$. Hence, the total restoring moment M is given by

$$M = 3\left(\frac{F_w\phi}{3}\right)r = F_w\phi r \tag{20.30}$$

and substitution for ϕ from Eq. (20.29) and equating the moment to the inertia torque give

$$M = -F_w\frac{r^2\theta}{l} = I_o\alpha \tag{20.31}$$

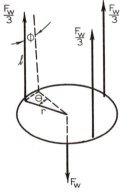

Fig. 20.16 Trifilar suspension.

where I_o is the moment of inertia of the platform and α is its angular acceleration. From Eq. (20.31),

$$\frac{\theta}{\alpha} = -\frac{I_o l}{F_w r^2} \tag{20.32}$$

and from Eq. (20.22) we find the period of small oscillations to be

$$T = 2\pi \sqrt{\frac{I_o l}{F_w r^2}} \tag{20.33}$$

Example 20.5 A connecting rod weighing 60 N is placed on a horizontal circular platform which is suspended by three wires, each 1.5 m long, from a rigid support. The wires are equally spaced around the circumference of a circle of 125-mm radius concentric with the platform. With the connecting rod positioned so that its center of mass lies directly above the center of the platform, the platform makes 10 small angular oscillations in 30 s. Determine the moment of inertia of the connecting rod about the vertical axis through its center of mass. The unladen platform weighs 15 N and makes 10 small angular oscillations in 35 s.

Solution. From Eq. (20.33), the moment of inertia I_o is given by

$$I_o = \frac{T^2}{4\pi^2}\left(\frac{F_w r^2}{l}\right) \tag{a}$$

For the connecting rod and platform,

$$I_{cp} = I_c + I_p = \frac{(3)^2}{4\pi^2}\left[\frac{(15 + 60)(0.125)^2}{1.5}\right]$$

$$= 0.178 \text{ kg·m}^2 \tag{b}$$

For the platform alone,

$$I_p = \frac{(3.5)^2}{4\pi^2}\left[\frac{15(0.125)^2}{1.5}\right]$$

$$= 0.0485 \text{ kg·m}^2 \tag{c}$$

Thus, the moment of inertia of the connecting rod is given by

$$I_c = I_{cp} - I_p = 0.178 - 0.0485 = 0.13 \text{ kg·m}^2$$

20.3.6 Single-Mass–Spring System

Figure 20.17 shows a mass m suspended from a vertical spring having a stiffness k (force per unit extension). If the mass is displaced downward by an amount x, the additional spring force ΔF, tending to return the mass to its equilibrium position, will be given by

$$\Delta F = kx \tag{20.34}$$

From the free-body diagram, shown in Fig. 20.17, we get

$$(\Delta F + mg) - mg = -m\ddot{x}$$

or substitution for ΔF and rearrangement give

$$\frac{x}{\ddot{x}} = -\frac{m}{k} \tag{20.35}$$

Since the ratio of the acceleration to the displacement is a negative constant, we can use Eq. (20.19) to obtain the period of oscillations. Thus,

$$T = 2\pi\sqrt{\frac{m}{k}} \tag{20.36}$$

Referring again to Fig. 20.17, when the mass is in its static equilibrium position, the spring has already been extended by an amount given by

$$mg = \delta k$$

or
$$\delta = \frac{mg}{k} \tag{20.37}$$

This deflection δ is the static deflection of the spring, and substitution of Eq. (20.37) into Eq. (20.36) gives

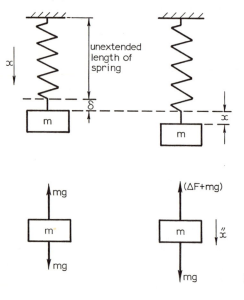

Fig. 20.17   Mass-spring system.

$$T = 2\pi \sqrt{\frac{\delta}{g}} \qquad (20.38)$$

It should be noted that in the preceding analysis the mass of the spring has been neglected.

Example 20.6 The pendulum shown in Fig. 20.18a is suspended from a fixed pivot at O. The pendulum consists of a bar B of weight 10 N and a block D of weight 60 N. The centers of mass C_1 and C_2 of B and D are at distances 150 mm and 375 mm from O. The radii of gyration of B and D each about its own center of mass are, respectively, 100 mm and 25 mm. A light spring is attached to the pendulum at a point P, 200 mm from O, and is anchored at a fixed point Q. When the pendulum is in static equilibrium, the line OC_1PC_2 is at 45° from the horizontal, and the angle OPQ is 90°. The spring stretches 25 mm for each 20 N of tension. Calculate the natural frequency of the pendulum for small oscillations about its equilibrium position.

Solution. If the pendulum is displaced through a small angle θ (Fig. 20.18b), the additional pull (restoring force) in the spring is given by

$$\Delta F = 200\theta\left(\frac{20}{25}\right) = 160\theta \text{ N} \qquad \text{(a)}$$

The moment of ΔF about O (the restoring moment due to the additional spring force) is therefore

$$\Delta M_f = 160\theta(0.2) = 32\theta \text{ N·m} \qquad \text{(counterclockwise)} \qquad \text{(b)}$$

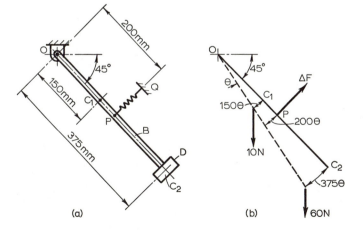

Fig. 20.18 Geometry for Example 20.6.

Due to the displacement of the pendulum there has been a reduction in the clockwise moments about O of the forces due to gravity. This change in moment is given by

$$\Delta M_w = 10\left(\frac{0.15\theta}{\sqrt{2}}\right) + 60\left(\frac{0.375\theta}{\sqrt{2}}\right) = 17\theta \text{ N·m} \qquad \text{(counterclockwise)}$$

$$(c)$$

The total restoring moment about O is therefore given by

$$M_o = \Delta M_f + \Delta M_w$$

$$= 32\theta + 17\theta$$

$$= 49\theta \text{ N·m} \qquad (d)$$

The moment of inertia of the pendulum about O is given by

$$I_o = m_B[k_{c1}^2 + (0.15)^2] + m_D[k_{c2}^2 + (0.375)^2]$$

$$= \frac{10}{9.81}[(0.1)^2 + (0.15)^2] + \frac{60}{9.81}[(0.025)^2 + (0.375)^2]$$

$$= 0.897 \text{ kg·m}^2 \qquad (e)$$

Finally,

$$M_o = -I_o\alpha \qquad (f)$$

or from Eq. (d),

$$49\theta = -0.897\alpha$$

and

$$\frac{\theta}{\alpha} = -0.183$$

Hence, from Eq. (20.22), we get

$$T = 2\pi\sqrt{-\frac{\theta}{\alpha}}$$

$$= 2\pi\sqrt{0.183} = 0.85 \text{ s}$$

20.4 Gyroscopic Effects

The preceding chapters dealing with rigid body dynamics have been concerned with the change in angular momentum of a rigid body, such as a disk or flywheel, caused by a moment acting in the plane of rotation, that is, when both the moment and the rotation are about parallel axes. The rotational equation of motion in this case was expressed as

$$M = \frac{d}{dt}(I\omega) = \frac{dp_0}{dt} \qquad (20.39)$$

where M and I are the external moment and the moment of inertia of the body, respectively, about an axis through either a fixed point or the center of mass and p_0 is the angular momentum about the same axis.

In the more general case, where the moment on the body acts in a plane other than the plane of rotation, the situation becomes a three-dimensional problem which is best handled by vector methods. One special three-dimensional case that can be handled rather simply, however, is when some rotationally symmetric body, such as a disk or flywheel, is rotating at a high speed about its axis of symmetry and is subjected to a moment which acts about an axis perpendicular to its axis of rotation. It is this special situation that is the subject of this section.

Equation (20.39) was developed for the situation where the motion of a rigid body is restricted to a single plane, such as the xy plane, for example, and in this case is the rotational equation of motion for the body about the z axis perpendicular to this plane. Clearly, Eq. (20.39) can also be applied to the other coordinate planes (xz and yz). Hence, for the general three-dimensional motion of a rigid body, three scalar rotational equations of motion exist, namely,

$$M_x = \frac{d}{dt}(I_x\omega_x) = \frac{dp_{ox}}{dt} \tag{20.40}$$

$$M_y = \frac{d}{dt}(I_y\omega_y) = \frac{dp_{oy}}{dt} \tag{20.41}$$

$$M_z = \frac{d}{dt}(I_z\omega_z) = \frac{dp_{oz}}{dt} \tag{20.42}$$

where the subscripts x, y, and z refer to the components of the various quantities about the x, y, and z axes, respectively. In vector notation, these equations can be written as

$$\mathbf{M} = \frac{d\mathbf{p}_o}{dt} \tag{20.43}$$

If the moment $\mathbf{M}$ is a constant, then Eq. (20.43) can be integrated to give

$$\mathbf{M}(t_2 - t_1) = \mathbf{p}_{o2} - \mathbf{p}_{o1}$$

or
$$\mathbf{M}\,\Delta t = \Delta\mathbf{p}_o \tag{20.44}$$

Figure 20.19a shows a disk which is rotating about its axis of symmetry. If a couple or moment Fl is applied in the xy plane (about the z axis), then from Eq. (20.44), the change in momentum about the z axis is given by

$$\Delta p_o = Fl\,\Delta t \tag{20.45}$$

Since we are considering situations where the angular momentum of the disk is much greater than the change in angular momentum Δp_o, then, from the

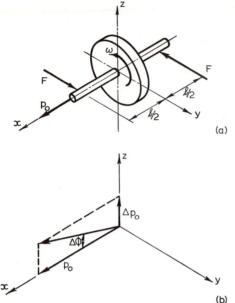

(a)

(b)　　**Fig. 20.19**　Spinning disk.

vector diagram in Fig. 20.19b, we see that

$$\Delta p_o = p_o\,\Delta\phi = I\omega\,\Delta\phi \tag{20.46}$$

where $I\omega$ is the initial angular momentum of the disk about the x axis and $\Delta\phi$ is the angle through which the angular momentum vector has rotated as a result of the applied momentum. Combining Eq. (20.44) with Eq. (20.46) gives

$$M\,\Delta t = I\omega\,\Delta\phi$$

or

$$M = I\omega\,\frac{\Delta\phi}{\Delta t} \tag{20.47}$$

In the limit, when $\Delta t \to 0$, Eq. (20.47) becomes

$$M = I\omega\frac{d\phi}{dt} = I\omega\Omega \tag{20.48}$$

where $\Omega = d\phi/dt$. Thus, the only effect of the applied moment is to give the angular momentum vector and, therefore, the axis of rotation of the disk (the spin axis), an angular velocity Ω in the xz plane, in a direction that rotates the angular momentum vector p_o toward the applied moment vector.

This rotation of the spin axis is called precession and is known as the gyroscopic effect. It is evident from Eq. (20.48) that when the disk spins rapidly, a large moment or couple is required to produce even a small precession, which explains the high stability of rapidly spinning bodies such as gyroscopes.

Figure 20.20 shows a simple method for the determination of the direction of precession. In this figure, the three arcs indicate the directions of the applied couple, the precession, and the spin and must be followed in the sense indicated by the arrows, which is the alphabetic order of the words couple, precession, and spin.

The behavior exhibited by a rapidly spinning disk comes as a surprise to most people unfamiliar with gyroscopic theory. For example, if the disk, suspended as shown in Fig. 20.21, is not spinning, it will fall due to its own weight F_w. However, if the disk is spinning rapidly, its axis of rotation will remain horizontal but precess about the vertical axis. This precession is caused by the moment about O of the weight F_w of the disk. Because of friction, however, the spin velocity will gradually decrease, and the disk will eventually fall.

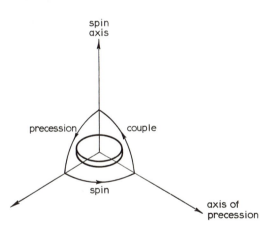

Fig. 20.20 Method for determination of direction of precession.

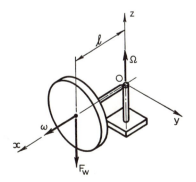

Fig. 20.21 Rapidly spinning disk.

Example 20.7   The ship's turbine rotor shown in Fig. 20.22a weighs 50 kN and has a radius of gyration about its spin axis of 250 mm. The rotor is supported horizontally in bearings at A and B and is spinning counterclockwise, as viewed from A toward B, at 200 s^{-1}. Determine the changes in the vertical components of the bearing reactions at A and B due to gyroscopic effects if the ship is making a 400-m radius turn to port at 10 m/s.

Solution.   From Eq. (20.48),

$$M = I\omega\Omega$$

or
$$1.7F = \left(\frac{50{,}000}{9.81}\right)(0.25)^2\left(\frac{200}{2\pi}\right)\left(\frac{10}{400}\right)$$

and
$$F = 149 \text{ N}$$

This force represents the change in the vertical reaction at each bearing due to the gyroscopic effect. The direction of the change for each bearing can be determined from the sketch shown in Fig. 20.22b and is indicated in Fig. 20.22a.

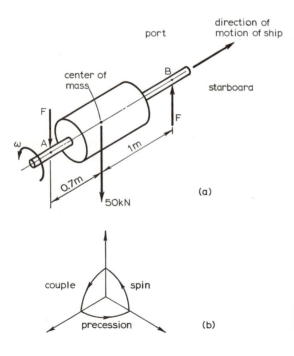

(a)

(b)

Fig. 20.22   Diagrams for Example 20.7.

Problems

20.1 A combined wheel and axle is connected through gears to a flywheel having a moment of inertia of 0.5 kg·m²; the gear ratio is such that the flywheel rotates twice as fast as the combined wheel and axle. The wheel and axle is used to raise a load of 500 N without acceleration by a falling weight of 200 N attached to a cord driving the wheel which has a diameter of 1.2 m. The load is suspended by a rope wrapped around the axle which has an effective radius of 100 mm. If the combined rotating parts of the wheel and axle have a weight of 400 N and a radius of gyration of 100 mm, find their angular acceleration and the tension in the rope to which the load is attached when the load is reduced to 400 N. Assume that the frictional resistance remains constant.

20.2 An automobile weighs 15 kN, and the moment of inertia of the wheels and rear axle is 7.5 kg·m² and of the engine parts is 0.25 kg·m². The gear ratios provided are 5.4, 7.3, 12.5, and 22.5 to 1, and the effective diameter of the road wheels is 0.7 m. If the engine torque is 45 N·m, find the maximum acceleration of the automobile in each gear.

20.3 Three masses A, B, and C weigh, respectively, 90 N, 80 N, and 130 N and rotate in the same plane at radii of 100 mm, 125 mm, and 50 mm, respectively. The angular positions of B and C are 60° and 135°, respectively, from A. Find the position and weight of a mass D at a radius of 150 mm to balance the system.

20.4 Five masses A, B, C, D, and E rotate in the same plane at equal radii. A, B, and C weigh, respectively, 50 N, 25 N, and 40 N. The angular positions of B, C, D, and E, measured in the same direction from A, are 60°, 135°, 210°, and 270°. Find the weights of the masses D and E for complete balance.

20.5 A shaft carries three pulleys A, B, and C at distances apart of 0.7 m and 1.4 m. The pulleys are out of balance to the extent of 25 N, 20 N, and 24 N, respectively, at a radius of 25 mm in each case. The angular positions of the out-of-balance masses in pulleys B and C with respect to that in pulley A are 90° and 210°, respectively. Determine, in position and magnitude, the balance weights required in planes L and M midway between planes A and B and B and C, respectively. The radius of rotation of the balance weights is 125 mm.

20.6 Five disks A, B, C, D, and E are equally spaced along a rotating shaft in planes perpendicular to the shaft axis and in this order along the shaft, the distance between successive planes being 225 mm. Initially, disks B and E are perfectly balanced, but disks A, C, and D, each of which weighs 180 N, have their centers of mass at 2 mm, 2.75 mm, and 1.25 mm, respectively, from the axis of rotation. The angular positions of the mass centers of C and D relative to A are 90° and 210°, respectively, both angles being measured in the same sense. The shaft is to be put into complete balance by drilling a hole in disk B and a hole in disk E, each with its center at 125-

mm radius. Determine the weight of material to be removed from each disk and the angular positions of the centers of the holes relative to disk *A*.

20.7 A thin disk of mass *m* is supported by a thin wire which is fixed to a rigid support (Fig. P20.7). If the torsional stiffness of the wire is *k*, obtain an expression for the moment of inertia of the disk in terms of the period of torsional oscillations of the disk.

20.8 A 40-N flywheel is suspended as a quadrifilar pendulum by four thin wires, equally spaced around its circumference (Fig. P20.8). Determine the moment of inertia of the flywheel about its axis of rotation if the flywheel completes 90 torsional oscillations in 60 s.

20.9 Obtain an expression for the natural frequency of oscillation of the mercury in a U-tube manometer open at both ends if the column of liquid mercury is of length *l* and specific weight γ (Fig. P20.9).

20.10 Obtain an expression for the natural frequency of the spring, mass, and pulley system shown in Fig. P20.10. Assume that the pulley bearing is frictionless and that the pulley has negligible mass.

20.11 For the arrangement shown in Fig. P20.11, determine the period of small oscillations about the equilibrium position shown. The springs each have a stiffness *k*, and the mass of the supporting member is small compared with that of the mass *m*.

20.12 The ring shown in Fig. P20.12 has a radius *r* and a negligible thickness. It swings with a small amplitude about the knife edge at *O*. Obtain an expression for the period of oscillations.

20.13 A cylinder of weight 5 kN has a center of mass displaced toward one end so that it floats in seawater in the upright position shown in Fig. P20.13. Neglecting any motion of the water, determine the period of oscillations of the cylinder if it is depressed slightly and then released. The specific weight of seawater is 10 kN/m³

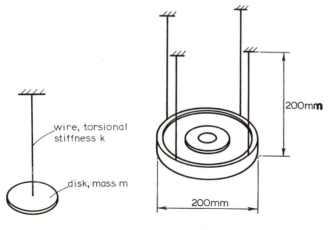

Fig. P20.7   Fig. P20.8

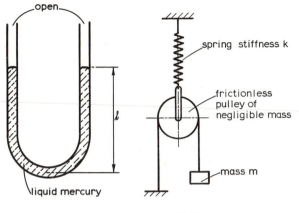

Fig. P20.9   Fig. P20.10

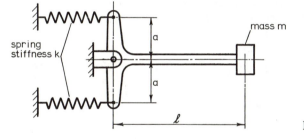

Fig. P20.11

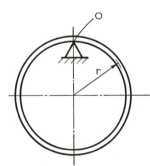

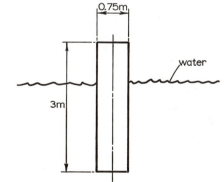

Fig. P20.12   Fig. P20.13

20.14   A ship is pitching through a total angle of 15°. The oscillations may be assumed to be simple harmonic, and the period is 32 s. The ship's turbine rotor weighs 50 kN, its radius of gyration is 440 mm, and it is rotating at 30 s^{-1}. Calculate the magnitude of the maximum gyroscopic couple caused by the rotor. If the rotation of the rotor is clockwise when viewed from aft, in which direction will the bow of the ship tend to turn when falling? What is the maximum angular acceleration to which the ship is subjected while pitching?

20.15   A 150-kW electric motor is installed in the cab of a large power shovel with its rotational axis horizontal. The rotor moment of inertia is $4.4 \text{ kg} \cdot \text{m}^2$, the rotor bearings are 1 m apart, and the motor shaft is connected to the driven member by use of a flexible coupling. If the rotational frequency of the rotor is 30 s^{-1} and the cab's maximum angular velocity about the vertical axis is 0.1 rad/s, what will be the additional bearing loads due to the gyroscopic effect?

20.16   A gas-turbine-powered racing car is being designed for use on a speedway were there are only left-hand turns. Considering the gyroscopic stabilizing effect, will there be a preferred and/or a least desirable orientation of the turbine rotor axis and its direction of rotation?

20.17   A rate-gyro is an arrangement which consists of a rotor spinning at high speed, mounted symmetrically in a frame (gimbal), and the frame, in turn, mounted symmetrically on a platform (Fig. P20.17). Initially the rotor spin axis is parallel to the platform. If the bearings are frictionless, then only the component about the x axis of the platform's rotation affects the motion of the gimbal. The moment of inertia of the rotor about its spin axis is $0.012 \text{ kg} \cdot \text{m}^2$, its rotational frequency is 330 s^{-1}, and the distance between the force transducers, which are attached between the gimbal and the

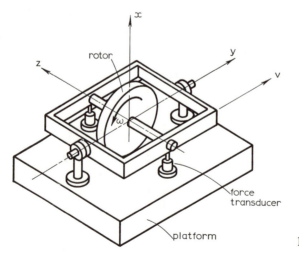

Fig. P20. 17

platform, is 40 mm. If the rate-gyro is to be designed to have a sensitivity (volts per newton) such that a platform angular velocity about the x axis of 0.1 mrad/s will provide an electrical signal of 0.01 V, determine the sensitivity of the transducer.

20.18   In rough seas, the transverse rolling motion of a ship can be reduced if a moment of the proper magnitude and phase is applied to the ship. One scheme employs a large, high-speed rotor mounted symmetrically in a gimbal such that its spin axis is vertical and the gimbal axis is athwartships (cross-wise) (Fig. P20.18). A second gyroscope assembly of similar attangement but much smaller in size is used to sense the rolling motion of the ship and to convert this into an electrical signal. This signal actuates a motor, thus causing the main gimbal to precess in response to the signal. This precession generates a moment that counteracts the roll of the ship. If the main rotor has a weight of 1 MN and a radius of gyration of 7 m and spins at 15 s⁻¹, determine the moment exerted on the ship's hull by the gyro if the motor turns the main gimbal at 40 mrad/s.

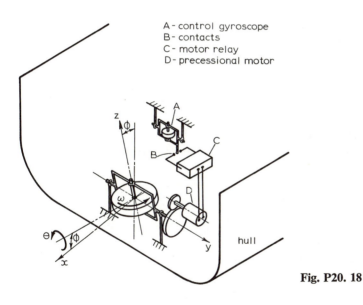

A - control gyroscope
B - contacts
C - motor relay
D - precessional motor

Fig. P20. 18

21

Project in Rocketry

21.1 Introduction

In this book no situations have been studied where the mass of a body changes during its motion, as is the case with rockets. Thus, this chapter on model rockets serves two purposes: (1) It illustrates how problems involving changes in mass can be handled; (2) it provides an introduction to a topic which can form an interesting project for the undergraduate engineering student. We shall first consider the basic equations governing the motion of a rocket in flight, and then we shall attempt to predict the altitude gained by a model rocket with an engine of specified performance.

21.2 Equation of Motion for a Rocket in Vertical Flight

A rocket engine produces gas molecules which are ejected with a high velocity; this causes a reaction which propels the rocket forward. To predict the motion of the rocket, it is necessary to obtain an equation which relates the forces acting on the rocket to its instantaneous acceleration. Forces are generally not applied to objects at a point but are distributed in their effect over a finite area. Under these circumstances, it is more useful to refer to a pressure; this is a force divided by the area over which the force is exerted. One effect of the ejection of gas molecules from a rocket engine is to produce a pressure at the rocket engine nozzle which helps to propel the rocket forward. The effect of this pressure is reduced to some extent by the pressure exerted by the atmosphere (atmospheric pressure). Thus, the effective pressure at the nozzle is found by subtracting the atmospheric pressure from the pressure of the exhaust gases. The force F_p resulting from this pressure is then found by multiplying the effective pressure p by the cross-sectional area A of the rocket nozzle.

Figure 21.1 shows all the externally applied forces acting on the rocket when it is in vertical flight, where F_d is the force due to air resistance (drag), F_w is the weight of the rocket and fuel, and F_p is the force due to the nozzle pressure.

The form of Newton's second law of motion employed in earlier chapters holds only for constant mass systems. Since fuel is consumed during the flight of the rocket, the total mass of the rocket and remaining fuel is changing, and the momentum form of the equation expressing Newton's second law must be used. Thus,

$$F = ma = m\frac{dv}{dt} = \frac{d(mv)}{dt} = \lim_{\Delta t \to 0}\frac{\Delta(mv)}{\Delta t} \tag{21.1}$$

Figure 21.2a shows the rocket and remaining fuel of mass m moving with a velocity v at time t. At time $t + \Delta t$, some of the fuel has been ejected, and the total mass has been divided into two portions (Fig. 21.2b): a portion of mass $m - \Delta m$ which is the rocket and remaining fuel moving with a velocity $v + \Delta v$ and the ejected fuel of mass Δm moving with a velocity v_f. If F_n is the

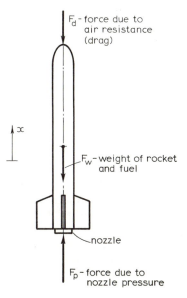

F_d - force due to air resistance (drag)

x

F_w - weight of rocket and fuel

nozzle

F_p - force due to nozzle pressure

Fig. 21.1   External forces acting on a rocket in vertical flight.

reaction force between the rocket and the ejected fuel, the equation of motion for the rocket and remaining fuel is

$$-F_d - (m - \Delta m)g + pA + F_n = \lim_{\Delta t \to 0} \left[\frac{(m - \Delta m)(v + \Delta v) - mv}{\Delta t} \right] \quad (21.2)$$

while the equation of motion for the ejected fuel is

$$-F_n - \Delta mg = \lim_{\Delta t \to 0} \left(\frac{\Delta m v_f}{\Delta t} \right) \quad (21.3)$$

Adding Eqs. (21.2) and (21.3) gives

$$-F_d - mg + pA = \lim_{\Delta t \to 0} \left(v_f \frac{\Delta m}{\Delta t} + m \frac{\Delta v}{\Delta t} - v \frac{\Delta m}{\Delta t} \right)$$

or

$$pA - (v_f - v)\frac{dm}{dt} - mg - F_d = m\frac{dv}{dt} \quad (21.4)$$

The sum of the first two terms on the left-hand side of this equation are defined as the thrust F_t of the rocket engine, and therefore,

$$F_t - mg - F_d = m\frac{dv}{dt} \quad (21.5)$$

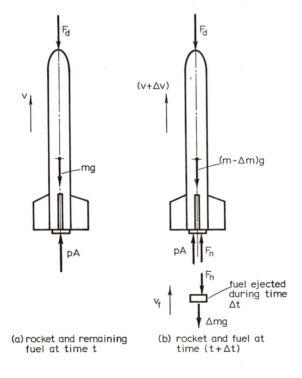

(a) rocket and remaining fuel at time t

(b) rocket and fuel at time (t+Δt)

Fig. 21.2 Change in mass of rocket and remaining fuel during flight; F_n = reaction force between rocket and ejected fuel, v = rocket velocity, and v_f = velocity of ejected fuel.

21.3 Engine Performance

In the previous section, the thrust of the rocket was defined as

$$F_t \;=\; pA \;+\; \left(-v_e \frac{dm}{dt}\right) \tag{21.6}$$

$$\underset{\substack{\text{total} \\ \text{thrust}}}{\downarrow} \quad \underset{\substack{\text{pressure} \\ \text{thrust}}}{\downarrow} \quad \underset{\substack{\text{momentum} \\ \text{thrust}}}{\downarrow}$$

where v_e is equal to $v_f - v$ and is the effective velocity of the exhaust gases. The term pA is referred to as the pressure thrust, while the remaining term is referred to as the momentum thrust; the negative sign is necessary because the mass is decreasing. The total trust F_t is obtained experimentally, for a particular engine, from measurements made on a static test stand. It is found that the pressure thrust is usually small compared with the momentum thrust and can be neglected.

In analyses of rocket engines, it is usual to express performance in terms of the specific impulse I_s, which is defined as the total thrust divided by the rate at which fuel is consumed. Thus,

$$I_s = \frac{F_t}{-g(dm/dt)} \tag{21.7}$$

If the pressure thrust pA is neglected, then

$$F_t = -\dot{m}v_e \tag{21.8}$$

and thus Eq. (21.7) can be written as

$$I_s = \frac{v_e}{g} \tag{21.9}$$

Often, the total impulse I_t of the engine is used and is given by

$$I_t = I_s F_{wf} \tag{21.10}$$

where F_{wf} is the initial weight of the fuel.

21.4 Aerodynamic Forces

One of the terms in the equation of motion for the rocket [Eq. (21.5)] was the air resistance or drag force F_d. This is the aerodynamic force due to the resistance to vertical motion of the rocket provided by the atmosphere. However, Fig. 21.3 shows the forces acting on the rocket when its axis is inclined at some angle of attack α to the velocity vector. In this case, the total air resistance F_r is made up of two components: One is the lift force F_l, which is perpendicular to the velocity vector, and the other is the drag force F_d, which is parallel to

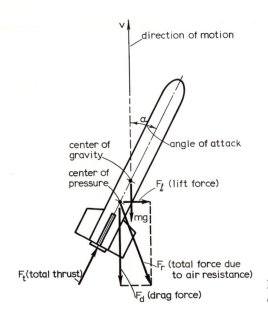

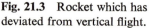

Fig. 21.3 Rocket which has deviated from vertical flight.

the velocity vector. These force components can be considered to act at a point in the rocket known as the center of pressure (Fig. 21.3). The total weight of the rocket can also be considered to act at a point: the center of gravity. In general, the lift and drag forces increase as the angle of attack of a rocket increases.

An important consideration in rocket design is its aerodynamic stability while passing through the atmosphere. A stable body is one that tends to return to its original orientation after being disturbed by an external force, while an unstable body is one that departs further from its original orientation when disturbed. Figure 21.4 shows examples of rockets in stable and unstable situations. In Fig. 21.4a, the center of pressure lies ahead of the center of gravity, and the rocket is unstable. In Fig. 21.4b, the center of pressure lies behind the center of gravity, and the rocket is stable. If the angle of attack α of the rocket is increased from zero, both the lift and drag forces will also increase. In the unstable case, this increase in lift and drag will cause the rocket to rotate about the center of gravity in such a way that α increases further. In the stable case, the increase in lift and drag will cause the rocket to rotate about the center of gravity in such a way that α decreases, causing the rocket to return to its initial orientation. To minimize the drag force, rockets usually take the form of long cylindrical shapes. Unfortunately, with a plain cylinder, the center of pressure lies ahead of the center of gravity, and this shape is therefore aerodynamically unstable. With the addition of suitable tail fins, the drag forces on the tail end of the rocket are increased to such an extent that

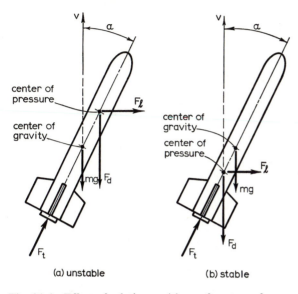

(a) unstable (b) stable

Fig. 21.4 Effect of relative positions of centers of pressure and gravity.

the center of pressure lies behind the center of gravity. Thus, the purpose of tail fins on a rocket is to provide aerodynamic stability.

At high altitudes, where the air density is very low, fins are not very effective because the air resistance becomes negligible. Thus, on multistage full-scale rockets, fins are generally restricted to the lower stages. For the lower stages of very large rockets, it becomes more practical to use engines that swivel and small steering rockets to obtain stability. For the upper stages of these rockets, spin stabilization is frequently used, taking advantage of the gyroscopic effect.

It is usual to express the drag force F_d in terms of the nondimensional drag coefficient C_d, defined as

$$C_d = \frac{F_d}{\frac{1}{2}\rho v^2 A_s} \qquad (21.11)$$

where ρ is the density of the air (mass per unit volume), v is the velocity of the rocket, and A_s is its cross-sectional area. The experimental determination of the drag coefficient will now be described.

21.5 Wind-Tunnel Tests

To determine the drag coefficient for a model rocket, it is necessary to perform measurements in a wind tunnel. A sketch of a typical wind-tunnel setup is given in Fig. 21.5. Facilities are provided for varying the speed of the air passing through the test section of the tunnel, and the air speed is indicated by a mercury manometer. A dynamometer, on which the rocket can be mounted, is capable of measuring the drag force. It should be noted that the force on the dynamometer stand will also be recorded by the dynamometer, and thus it is necessary to measure the drag force with and without the rocket in position. The drag force on the rocket can then be found by subtraction of these two readings.

Figure 21.6 presents the results of a typical test where the drag force F_d is plotted against the square of the air speed v^2 for a range of speeds from 13 to 31 m/s. The results show that F_d is proportional to v^2, or, in other words, F_d/v^2 is a constant.

Since F_d/v^2, ρ, and A_s were all constant during the test, the drag coefficient C_d will also be a constant for the range of conditions investigated. Thus, the drag coefficient can be calculated for each test conducted and the results averaged to give the mean value of C_d and the confidence limits for the mean.

Prior to performing this calculation, it is necessary to determine the density of the air in the wind tunnel, and for this purpose the ambient temperature and atmospheric pressure should be recorded. The standard density of air can then be corrected for the conditions of the test. In the experiment described here, the mean drag coefficient C_d was 0.773 with 95% confidence limits of ± 0.067.

21.6 Computation of Maximum Altitude

For model rockets, the drag coefficient C_d can be regarded as a constant, and thus substitution of Eq. (21.11) into Eq. (21.5) gives, after rearrangement,

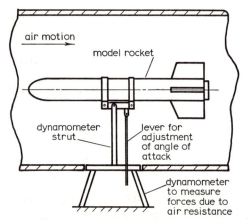

air motion

model rocket

dynamometer strut

lever for adjustment of angle of attack

dynamometer to measure forces due to air resistance

Fig. 21.5 Wind-tunnel setup for test of a model rocket.

$$\frac{dv}{dt} = \frac{F_t}{m} - \frac{\rho}{2m} A_s C_d v^2 - g \tag{21.12}$$

It should be remembered that the mass m refers to the rocket and remaining (unburnt) fuel and will therefore vary during flight. If m_o is the mass of the rocket and F_{wf} is the initial weight of the fuel, then the initial value of m is $(m_o + F_{wf})/g$, and the final value, after burnout (when all fuel has been consumed), is m_o.

After burnout, the thrust term F_t/m in Eq. (21.12) will become zero, and the rocket will continue in free flight until it reaches it maximum altitude, when its velocity becomes zero.

It is not possible to solve Eq. (21.12) by analytical means, but a numerical solution can readily be obtained with the aid of a programmable calculator. The simplest procedure is to divide the total flight time into small intervals Δt and, starting from the ignition of the engine, to calculate the velocity and altitude of the rocket after each successive interval of time. For this purpose, Eq. (21.12) can be rewritten as

$$\Delta v = \left(\frac{F_t}{m} - \frac{\rho}{2m} A_s C_d v^2 - g \right) \Delta t \tag{21.13}$$

which gives the increase in velocity during each time interval. The vertical

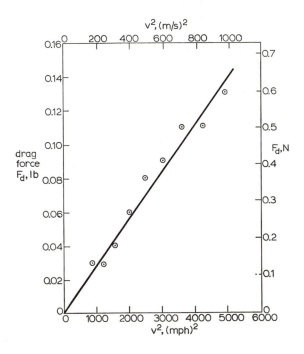

Fig. 21.6 Drag force on Centurion model rocket.

distance traveled during each time interval will be given by

$$\Delta x = v \, \Delta t \tag{21.14}$$

where x is the altitude of the rocket. The change Δm in the mass of the rocket and fuel can be obtained from Eq. (21.8). Thus,

$$\Delta m = -\left(\frac{F_t}{v_e}\right) \Delta t \tag{21.15}$$

It should be noted that the fuel is assumed to be consumed at a constant rate, giving a constant thrust F_t.

In a computation such as this, the greatest accuracy is obtained when the shortest time intervals are chosen. For the purposes of illustration, a hand computation is presented below where the total burn time is divided into only 10 intervals. The use of a programmable calculator would allow shorter time intervals to be employed, resulting in a more accurate result.

21.6.1   Hand Computation

The following example is for a Centurion model rocket with a C6–5 engine, and the following data are applicable:

$$\text{rocket weight} = 0.640 \text{ N}$$
$$\text{engine weight (including fuel)} = 0.253 \text{ N}$$
$$\text{propellant (fuel) weight } F_{wf} = 0.122 \text{ N}$$
$$\text{burning time } t_b = 1.7 \text{ s}$$
$$\text{total impulse } I_t = 10 \text{ N·s}$$
$$\text{average total thrust } F_t = 5.87 \text{ N}$$
$$\text{drag coefficient } C_d = 0.773$$
$$\text{rocket diameter } d = 0.042 \text{ m}$$
$$\text{mass density of air } \rho = 1.2 \text{ kg/m}^3 \text{ (N·s}^2/\text{m}^4)$$
$$\text{time from ignition to parachute ejection } t_p = 6.7 \text{ s}$$

The specific impulse I_s is obtained by dividing the total impulse I_t by the fuel weight F_{wf}. Thus,

$$I_s = \frac{I_t}{F_{wf}} = \frac{10}{0.122} = 82 \text{ s}$$

The relative velocity v_e of the ejected fuel is given by

$$v_e = I_s g = 82(9.81) = 804 \text{ m/s}$$

Hence, from Eq. (21.15),

$$-\Delta m = \left(\frac{F_t}{v_e}\right) \Delta t = \frac{5.87}{804} \Delta t$$

or
$$-\Delta m = 7.3 \times 10^{-3} \Delta t$$

The initial value of m is given by

$$m_i = m_o + \frac{F_{wf}}{g}$$

$$= \frac{0.893}{9.81} = 0.091 \text{ kg}$$

The cross-sectional area of the rocket A_s is

$$A_s = \frac{\pi d^2}{4} = \frac{\pi (0.042)^2}{4} = 0.00139 \text{ m}^2$$

Finally, the factor

$$\frac{\rho}{2} A_s C_d = \frac{1.2}{2} (0.00139)(0.773) = 6.43 \times 10^{-4} \text{ N} \cdot \text{s}^2/\text{m}^2$$

Table 21.1 presents the hand computation, and the results are presented in graphical form in Fig. 21.7, where it can be seen that the maximum altitude is reached 5 s after ignition. The parachute opens after the rocket has reached maximum altitude and when the rocket is at an altitude of about 125 m.

It is interesting to build and test a model rocket and compare the actual altitude gained with that predicted by theory and computation.

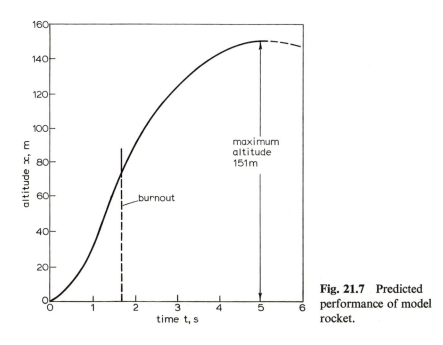

Fig. 21.7 Predicted performance of model rocket.

Table 21.1 Hand Computation: $C_d = 0.773$, $\Delta t = 0.17$ s, $\Delta m = -7.3 \times 10^{-3}$, $\Delta t = -1.24 \times 10^{-3}$ kg.

	1	2	3	4	5	6	7	8	9
(1) v	0	9.30	18.6	27.7	36.5	44.7	52.2	58.9	64.8
(2) v^2	0	86.5	346	767	1332	1998	2725	3469	4199
(3) $m = m_i + \Delta m$	0.091	0.090	0.089	0.087	0.086	0.085	0.084	0.082	0.081
(4) $v^2/m = (2)/(3)$	0	961	3909	8788	15,488	23,505	32,440	42,304	51,788
(5) $(\rho/2)A_sC_a \times (4) = 6.43 \times 10^{-4} \times (4)$	0	0.618	2.51	5.65	9.56	15.1	20.8	27.2	33.3
(6) $(F_t/m) - 9.81 = [5.87/(3)] - 9.81$	54.7	55.4	56.1	57.7	57.7	59.2	60.1	61.8	62.6
(7) $\Delta v = [(6) - (5)] \times 0.17$	9.30	9.31	9.12	8.84	8.18	7.50	6.67	5.88	4.98
(8) $v_2 = (1) + (7)$	9.30	18.6	27.7	36.5	44.7	52.2	58.9	64.8	69.8
(9) $\Delta x = (8) \times 0.17$	1.58	3.16	4.71	6.21	7.60	8.87	10.0	11.0	11.9
(10) $x_2 = (9) + (10)$	1.58	4.74	9.45	15.7	23.3	32.2	42.2	53.2	65.1

	10	11	12[a]	13	14	15	16	17	18
		Burnout							
(1) v	69.8	73.9	64.7	49.8	39.6	31.9	25.7	20.5	16.0
(2) v^2	4872	5461	4186	2480	1568	1018	660	420	256
(3) $m = m_i + \Delta m$	0.080	0.079	0.079	0.079	0.079	0.079	0.079	0.079	0.079
(4) $v^2/m = (2)/(3)$	60,900	69,129	52,988	31,393	19,850	12,881	8361	5320	3240

(5) $6.43 \times 10^{-4} \times (4)$	39.1	44.4	34.1	20.2	12.8	8.28	5.38	3.42	2.08
(6) $(F_t/m) - 9.81$	63.5	-9.81	-9.81	-9.81	-9.81	-9.81	-9.81	-9.81	-9.81
(7) $\Delta v = [(6) - (5)] \times 0.17^a$	4.15	-9.22	-14.9	-10.2	-7.67	-6.15	-5.16	-4.50	-4.04
(8) $v_2 = (1) + (7)$	73.9	64.7	49.8	39.6	31.9	25.7	20.5	16.0	12.0
(9) $\Delta x = (8) \times 0.17^a$	12.6	11.0	16.9	13.5	10.8	8.75	6.98	5.44	4.06
(10) $x_2 = (9) + (10)$	77.7	78.7	95.3	109	120	129	136	141	145

	19	20	21	22
(1) v	12.0	8.27	4.74	1.34
(2) v^2	144	68.4	22.5	1.80
(3) m	0.079	0.079	0.079	0.079
(4) $v^2/m = (2)/(3)$	1823	865	284	22.7
(5) $6.43 \times 10^{-4} \times (4)$	1.17	0.557	0.183	0.015
(6) -9.81	-9.81	-9.81	-9.81	-9.81
(7) $\Delta v = [(6) - (5)] \times 0.34$	-3.73	-3.52	-3.40	-3.34
(8) $v_2 = (1) + (7)$	8.27	4.74	1.34	-2.00
(9) $\Delta x = (8) \times 0.34$	2.81	1.61	0.456	-0.681
(10) $x_2 = (9) + (10)$	148	150	151	150

[a]After burnout Δt was changed to 0.34 s.

Appendixes

Appendix I: SI System of Units (A Selection of Basic and Derived Quantities).

Quantity	Symbol	Definition	Name of Unit	International Symbol for Unit
Length	l	See Chapter 1	Metre	m
Breadth	b			
Height	h			
Thickness	d, δ			
Radius	r			
Diameter	d			
Length of path	s			
Angle (plane angle)	$\alpha, \beta, \gamma, \theta, \phi,$ etc.	The angle between two half-lines terminating at the same point is defined as the ratio of the arc cut out on a circle (with its center at that point) to the radius of the circle	Radian	rad
			Degree	°
Area	A		Square metre	m²
Volume	V		Cubic metre	m³
Time	t	See Chapter 1	Second	s
Angular velocity	ω	$\omega = d\phi/dt$	Radian per second	rad/s
Angular acceleration	α	$\alpha = d\omega/dt$	Radian per second squared	rad/s²
Velocity	v	$v = ds/dt$	Metre per second	m/s

Quantity	Symbol	Definition	Unit name	Unit symbol
Acceleration	a	$a = dv/dt$	Metre per second squared	m/s²
Acceleration of free-fall	g	$g = 9.806\ 65$ m/s²		
Periodic time	T	Time of one cycle	Second	s
Frequency	f	$f = 1/T$	Hertz	Hz
Rotational frequency	n	Number of revolutions divided by time	Reciprocal second	s⁻¹
Angular frequency	ω	$\omega = 2\pi f$	Reciprocal second	s⁻¹
Mass	m	See Chapter 1	Kilogram Tonne (1 t = 1000 kg)	kg t
Density (mass density)	ρ	Mass divided by volume	Kilogram per cubic metre	kg/m³
Momentum	p	Product of mass and velocity	Kilogram-metre per second	kg·m/s
Moment of momentum	p_0	The moment of momentum of a particle about a point is equal to the vector product of the radius vector from this point to the particle and the momentum of the particle	Kilogram-metre squared per second	kg·m²/s
Angular momentum	p_0			

(continued)

Appendix I (cont.)

Quantity	Symbol	Definition	Name of Unit	International Symbol for Unit
Force	F		Newton (1 N is that force which when applied to a body having a mass of 1 kg gives it an acceleration of $1 \ m/s^2$)	N
Weight	W	The weight of a body is that force which when applied to the body would give it an acceleration equal to the local acceleration of free-fall	Newton	N

Appendix II: Properties of Plane Figures.

Figure		Location of Centroid	Second Moment of Area
Arc segment	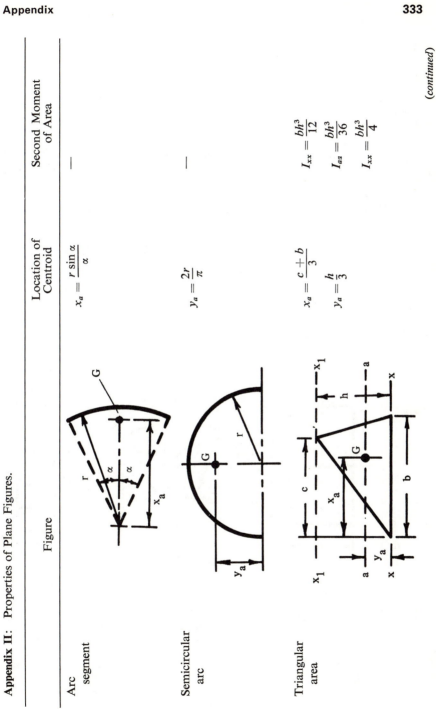	$x_a = \dfrac{r \sin \alpha}{\alpha}$	—
Semicircular arc		$y_a = \dfrac{2r}{\pi}$	—
Triangular area		$x_a = \dfrac{c + b}{3}$ $y_a = \dfrac{h}{3}$	$I_{xx} = \dfrac{bh^3}{12}$ $I_{aa} = \dfrac{bh^3}{36}$ $I_{xx} = \dfrac{bh^3}{4}$

(continued)

Appendix II (cont.)

Figure	Location of Centroid	Second Moment of Area
Rectangular area	—	$I_{xx} = \dfrac{bh^3}{3}$ $I_{aa} = \dfrac{bh^3}{12}$
Circular sector	$x_a = \dfrac{2}{3} \dfrac{r \sin \alpha}{\alpha}$	$I_{xx} = \dfrac{r^4}{4}\left(\alpha - \dfrac{\sin 2\alpha}{2}\right)$ $I_{yy} = \dfrac{r^4}{4}\left(\alpha + \dfrac{\sin 2\alpha}{2}\right)$
Quarter circular area	$x_a = y_a = \dfrac{4r}{3\pi}$	$I_{xx} = I_{yy} = \dfrac{\pi r^4}{16}$

Elliptical
quadrant
$(A = \pi ab/4)$

$x_a = \dfrac{4a}{3\pi}$

$y_a = \dfrac{4b}{3\pi}$

$I_{xx} = \dfrac{\pi ab^3}{16}$

$I_{yy} = \dfrac{\pi a^3 b}{16}$

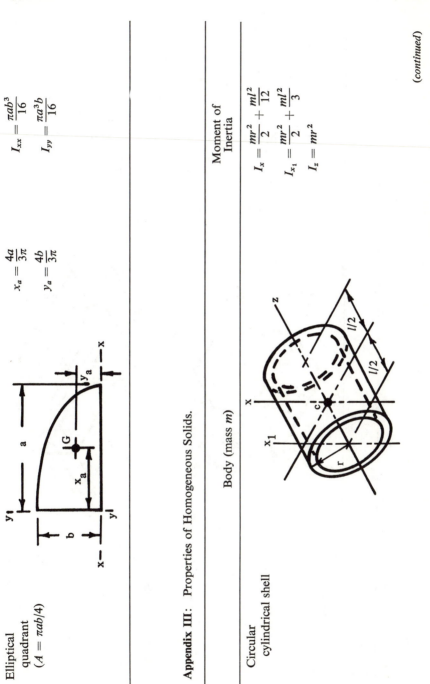

Appendix III: Properties of Homogeneous Solids.

Body (mass m)	Moment of Inertia
Circular cylindrical shell	$I_x = \dfrac{mr^2}{2} + \dfrac{ml^2}{12}$ $I_{x_1} = \dfrac{mr^2}{2} + \dfrac{ml^2}{3}$ $I_z = mr^2$

(continued)

Appendix III (cont.)

Body (mass m)	Moment of Inertia

Circular cylinder

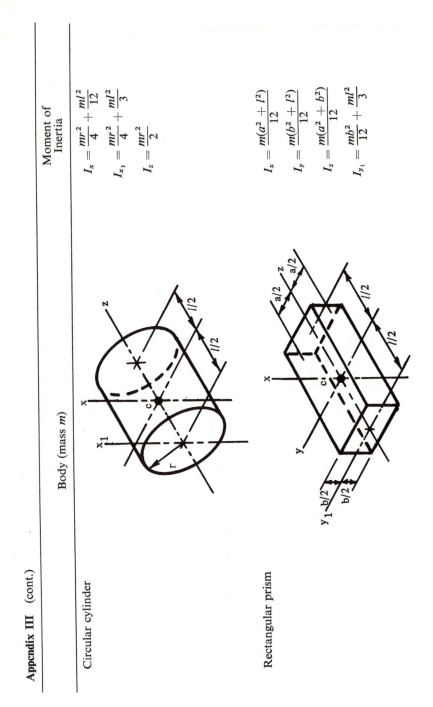

$$I_x = \frac{mr^2}{4} + \frac{ml^2}{12}$$

$$I_{x_1} = \frac{mr^2}{4} + \frac{ml^2}{3}$$

$$I_z = \frac{mr^2}{2}$$

Rectangular prism

$$I_x = \frac{m(a^2 + l^2)}{12}$$

$$I_y = \frac{m(b^2 + l^2)}{12}$$

$$I_z = \frac{m(a^2 + b^2)}{12}$$

$$I_{y_1} = \frac{mb^2}{12} + \frac{ml^2}{3}$$

Spherical shell

$$I_z = \frac{2}{3}mr^2$$

Sphere

$$I_z = \frac{2}{5}mr^2$$

Right circular cone

$$I_y = \frac{3}{20}mr^2 + \frac{3}{5}mh^2$$

$$I_{y_1} = \frac{3}{20}mr^2 + \frac{1}{10}mh^2$$

$$I_z = \frac{3}{10}mr^2$$

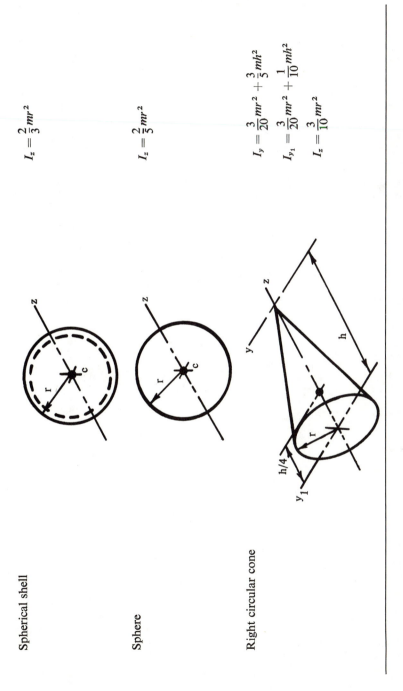

Nomenclature

A	Area; area of a surface; area of a cross section
A_s	Cross-sectional area of a rocket
a	Acceleration
a_{av}	Average acceleration
a_n	Normal component of acceleration
a_r	Radial component of acceleration
a_t	Tangential component of acceleration
a_x	Component of acceleration parallel to x axis
a_y	Component of acceleration parallel to y axis
a_{BA}^r	Radial component of acceleration of point B relative to point A (both points being on the same rigid body)
a_{BA}^t	Tangential (or angular) component of acceleration of point B relative to point A (both points being on the same rigid body)
a_θ	Angular component of acceleration
C_d	Drag coefficient
d	Distance; distance between two parallel axes; diameter
e	Coefficient of restitution
F	Force
F_A, F_B, etc.	Forces applied at points A, B, etc.
F_{AB}, F_{AC}, etc.	Forces in members AB, AC, etc.
F_d	Drag force
F_e	Equilibrant force
F_f	Frictional force; tangential force
F_l	Lift force

339

F_n	Normal force; reaction force
F_r	Resultant force; force due to air resistance
F_s	Shear force
F_t	Tension force; thrust force of rocket engine
F_{t1}	Tension force in tight side of belt or rope
F_{t2}	Tension force in slack side of belt or rope
F_w	Weight of an object
F_{wf}	Initial weight of fuel in a rocket
F_x	Component of force parallel to x axis
F_y	Component of force parallel to y axis
$\hat{F}$	Impulsive force
G	Universal gravitational constant
g	Acceleration due to gravity
h	Height; depth of submerged point from surface of liquid
h_a	Depth of centroid of area of submerged surface, measured from surface of liquid
I	Moment of inertia; second moment of area
I_{aa}	Second moment of area of a plane area about the $a\text{-}a$ axis
I_c	Moment of inertia of a body about an axis through its center of mass
I_e	Equivalent moment of inertia of a geared system
I_o	Moment of inertia of a body about an axis through the fixed point O
I_s	Specific impulse of rocket engine
I_t	Total impulse of rocket engine
I_x, I_y, etc.	Moment of inertia of a body about x, y axes, etc.
I_{xx}, I_{yy}, etc.	Second moment of area of a plane area about the x axis, y axis, etc.
I_{xy}	Product of inertia of a plane area with respect to the x, y axes.
k	Stiffness of spring (force per unit extension or compression); radius of gyration of a body
k_c	Radius of gyration of a body about an axis through its center of mass
k_o	Radius of gyration of a body about an axis through point O
M	Moment; torque; bending moment
M_A, M_B, etc.	Moment about points A, B, etc.
M_o	Moment of external forces about an axis through point O

$\hat{M}$	Impulsive moment
m	Mass
m_e	Mass of earth
m_l	Mass per unit length
m_o	Mass of rocket
N	Ratio of rotational motions
n	Frequency of vibration; rotational frequency (number of revolutions per unit time)
P	Power
p	Pressure (normal load per unit area); pitch of screw thread
p_{av}	Average pressure
p_c	Angular momentum about axis through center of mass
p_g	Gage pressure
p_{ga}	Gage pressure at centroid of a submerged surface
p_o	Angular momentum about axis through point O; pressure at surface of liquid
p_o'	Angular momentum after impact
r	Distance from origin of coordinate system; moment arm; radius
r_c	Distance from center of mass
r_o	Distance from point O
s	Distance
T	Kinetic energy; periodic time of a vibratory motion (time for one cycle)
T_r	Kinetic energy of a body due to rotation about its center of mass
T_t	Kinetic energy of a body due to translational motion (of its center of mass)
T'	Kinetic energy after impact
t	Time
t_b	Burning time for rocket engine
t_p	Time from ignition of rocket engine to parachute ejection
V	Potential energy; volume
v	Velocity
v_{av}	Average velocity
v_e	Effective velocity of exhaust gases
v_f	Velocity of fuel
v_r	Radial component of velocity
v_t	Tangential component of velocity
v_x	Component of velocity parallel to x axis

v_y	Component of velocity parallel to y axis
v'	Velocity after impact
v_θ	Angular component of velocity
W	Work
W_c	Work done by conservative forces
W_f	Work done by frictional forces
W_n	Work done by nonconservative forces
W_r	Work done by moment causing rotation of rigid body
W_t	Work done due to translation of (center of mass of) rigid body
w	Load per unit beam length of distributed load
x_a	x coordinate of centroid of area
x_c	x coordinate of center of mass
x_g	Distance to equivalent point load; x coordinate of center of gravity
x_p	x coordinate of center of pressure
x_v	x coordinate of center of volume
y_a	y coordinate of centroid of area
y_c	y coordinate of center of mass
y_g	y coordinate of center of gravity
y_p	y coordinate of center of pressure
y_v	y coordinate of center of volume
α	Angular acceleration; helix angle of a screw thread; angle of attack of rocket
β	Half-angle of V-thread; half-angle of V-belt; angle of friction
β_d	Angle of dynamic friction
β_s	Angle of static friction
γ	Specific weight (weight per unit volume)
δ	Deflection of spring under static load
θ	Angular position
θ_w	Angle of wrap (angle subtended by arc of contact between belt or rope and pulley)
μ	Coefficient of friction
μ_d	Coefficient of dynamic friction
μ_e	Equivalent coefficient of friction
μ_s	Coefficient of static friction
ρ	Density; radius of curvature of curved path
ρ_m	Density of mercury
Ω	Angular velocity of spin axis (rate of precession)
ω	Angular velocity
ω'	Angular velocity after impact

Index